Math Facts

Kids Need Them. Here's How to Teach Them.

DEVELOPING MATHEMATICAL THINKING INSTITUTE (DMTI)

Jonathan Brendefur, PHD

Sam Strother, MAE

Published by Developing Mathematical Thinking Institute
www.dmtinstitute.com

Copyright © 2021 by Jonathan Brendefur

All rights reserved.

No part of this book may be reproduced or transmitted in any form or by any means, electronic or mechanical, including photocopying, recording, or by any information storage or retrieval system, without the written permission of the authors, except where permitted by law.

Cover design by Chris Collins
Edited by Rachel Ann Horatia Beckett
Interior page design and composition by Cecile Kaufman

ISBN: 978-1-7373290-0-8

Printed in the United States of America

Contents

Introduction 1
Teaching Mathematics 1
The DMT Instructional Model 3
Math Fluency 6
    The Fluency Debate 7
    Can Drilling Math Facts Be Effective? 8
The Research and History of Fact Fluency 9
    Why Don't Schools Use Research-based Fluency Development Strategies? 10

Fluency Introduction: Addition and Subtraction 13
Spatial Reasoning Skills Aid in Developing Fluency 14
    The Importance of Game Play 15
Addition and Subtraction Fluency and Flexibility (0–20) 15
    Composing and Decomposing 16
    TEMPLATE: Composing a Number 17
    Subitizing 17
    TEMPLATE: Dot Patterns 18
Addition and Subtraction Strategies 19
    Doubling 20
    Benchmark Strategy 23
More Advanced Strategies 26
    Place Value Strategy 26
    TEMPLATE: Place Value Strategy 30
    Compensation Strategy 31
Addition and Subtraction Strategy Summary 35

Addition and Subtraction Fluency and Flexibility with All Numbers 36

Fluency and Flexibility for Multi-digit Numbers 36

- Doubles 37
- Benchmark Strategy 38
- Place Value Strategy 38
- Compensation Strategy 39
- Doubles and Compensation Strategy 40

Fluency and Flexibility with Decimals and Fractions 40

- Summary 43
- TEMPLATE: Addition and subtraction strategy practice (multi-digit) 43
- TEMPLATE: Addition and subtraction strategy practice (decimals) 43
- TEMPLATE: Addition and subtraction strategy practice (fractions) 44
- TEMPLATE: Addition and subtraction strategy practice 44

Multiplication and Division Fluency and Flexibility (0–20) 45

Initial Ideas to Building Multiplicative Thinking 46

- Skip counting 46
- Area models 48
- TEMPLATE: Decomposing Numbers 2 to 10 Using an Open Area Model 53

Multiplication and Division Strategies 53

- Doubles Strategy 53
- TEMPLATE: Doubles Strategy 55
- Squares Strategy 58
- TEMPLATE: Squares Strategy 61
- Benchmark Strategy 62
- TEMPLATE: Benchmark Strategy 65
- Compensation 66
- TEMPLATE: Compensation Strategy 68
- Practice and Probes 69
- TEMPLATE: Multiplication Strategy Practice 69
- TEMPLATE: Strategy Cards 71
- TEMPLATE: Multiplication Matrix 72

Multiplication and Division Fluency and Flexibility (Multi-digit and Rational Numbers) 73

Fluency and Flexibility with Multi-digit Multiplication Using Integers 74

Fluency and Flexibility with Multi-digit Multiplication Using Non-integer Rational Numbers 81

Conclusion 84

Introduction

TEACHING MATHEMATICS

How do children learn best? National organizations and researchers are clear on how students learn and retain information, but typical math instruction is not in alignment. Why? There are several different reasons and we will highlight each and explain how to maximize our children's learning.

Students are typically taught math through lectures and asked only to reproduce knowledge that has been presented to them. A reliance on this approach alone ensures only a few will like, understand, remember, and use mathematics as an adult. How many times have we heard someone say, "I hate math!" "I'll never need math!" "I was never good in math!" We want to make sure the opposite occurs: "I love math!" "I use math all the time!" "I'm good at math!" To be good at math is to be able use it to solve problems, make sense of the world around us, and communicate effectively, especially in a data-filled society.

So why, then, do learning experiences for many math students not follow what researchers have long suggested to be most beneficial? There are five reasons for this misalignment.

Reason #1: Teaching for understanding is time consuming and difficult. When students are first learning something, it is important for them to understand it conceptually or to know how and why it works the way it does. By understanding the boundaries (or parameters) of an idea or method, a student builds a foundation. Then, balanced with the conceptual understanding, the student will practice a procedure or algorithm. Teaching that starts with the end product or the procedure will do more harm than good to a student's understanding of math in the long term. For example, if you were teaching how to add numbers using place value, there might be a temptation to teach the formal regrouping algorithm that involves "carrying" digits as the first and only method, focusing on these very specific procedural steps before students have had a chance to learn more expanded forms that illustrate what happens as you add each unit of place value. Teaching procedurally is easier because we know the steps and the answer. We must learn how to incorporate the conceptual knowledge before, and along with, the procedural knowledge.

```
 11
 678      600      70      8
+145     +100     +40     +5
 823      700     110     13
```

The place value method of addition and how it needs to be deconstructed to communicate it to students.

Reason #2: Teaching for understanding is foreign to most people. We ourselves were perhaps not taught how the mathematics works, or why or when we would use it, so we do not know any approach other than to provide examples. We are also afraid that if students are exploring the mathematics and ask a question we cannot answer, we shall become uncomfortable and have to revert to what we know — the procedures. We must be willing to be okay with this discomfort and learn the how, why, and when ourselves, as educators. Think of it as we also become explorers, with the student, admitting to the limits of our own knowledge, confidence and a sense of excitement and determination to understand mathematics better.

Reason #3: The organization of schools can constrain teaching. Teachers and parents must work together and support each other to create situations for students to understand the subject and then flexibly practice it. This is particularly difficult with mathematics as many parents in the United States lack a comfort level and knowledge base adequate to teach mathematics to children who may be struggling. This problem is likely the result of parents' own math learning experiences in their schooling, which perhaps focused on procedures and mechanical techniques over understanding and connections. Teachers and parents working collaboratively can help improve and modernize math learning for students.

Reason #4: Textbooks have not changed. Most math curricular resources still provide an example and then ask students to practice a rote procedure. It makes sense that if a student does not understand the math or use it practically in the world around them, then they will soon dislike it and struggle. We have the ability now with technology to be more flexible and can put students in situations to ensure they understand and know how to apply the mathematics.

Reason #5: A belief that only our advanced students can engage in critical thinking. If we only focus on procedural knowledge, then students who have the ability to remember isolated facts or can figure out patterns on their own, will succeed, or at least for a while. Interestingly, spatial reasoning is one of the best predictors of success in math. Spatial reasoning is commonly defined as the ability to mentally visualize how shapes and objects can be rotated or manipulated to make visual comparisons. You are using spatial reasoning skills when you decide how to pack a piece of luggage, move a piece of furniture through a doorway, or create a piece of artwork. Many of our students come to school great at manipulating space but then this is not explicitly taught or shown to be connected with the topic of number. Many of these students believe they are failures when, in fact, they are excellent problem-solvers, are ripe for understanding the math and could become very successful over time. We must focus on ensuring all students have a more robust set of situations. To do this will require a fundamental shift in many peoples' beliefs about students and their abilities. At the DMTI,[1] we often suggest teachers and parents focus on what scaffolds learning and supports children with what they may need to access math content, instead of having limited beliefs about their abilities if they have struggled in the past.

In this next section, we will explain how to develop mathematical thinking.

THE DMT INSTRUCTIONAL MODEL

As was mentioned in the introduction, students need to understand math through a balanced approach combining conceptual and procedural knowledge, enabling them to become fluent, but also to become excellent problem-solvers, making connections inside and outside of math classrooms and to use and communicate with math. To understand math means to know both why to do something and how. When we know why and how, we are able to flexibly use procedures and understand their interrelationships within the larger structure of mathematics. Knowing both why and how also extends our knowledge of how to use math to solve related problems in similar situations — in context and out (horizontal mathematizing), and then extend our knowledge to new ideas (vertical mathematizing).

---

[1] Developing Mathematical Institute (DMTI) is a math research-based education company that provides professional development with curricular resources and diagnostic assessments. Our mission is to work with administrators, teachers and students to create a partnership to improve mathematics teaching and learning.

Cognitively, our knowledge is structured with web-like and hierarchal connections. Each math topic or concept is a network of these representations. So, each idea in math can be thought of as a node with one or more connections. The greater the number of these connections and the stronger each individual connection is, the greater our understanding of this topic. And the more connections we have, the greater our ability to use this knowledge to solve new and more complex problems. Each web of knowledge is a schema for how we think about and use the mathematics. For example, imagine 3 × 4. What are all the ideas that come to mind? You might think, 12 is the product. But, what else? It is also 4 × 3. I can imagine a 3 × 4 rectangle or a 4 × 3 array. I know I could have a context, with, say, 4 flowers with 3 petals each or 3 flowers with 4 petals each. I know I could count by 3s, four times, to get to 12 or by 4s, three times, to get to 12. I also know I could decompose the 4 into 2 x 2, so 2 × 2 × 3 is 12. The stronger your schema on this topic the more connections you will make and the better your ability will be to know and use the mathematics.

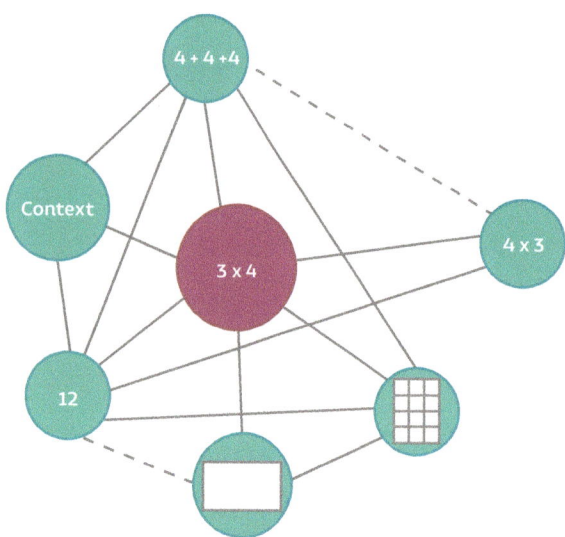

Example of a schema map for 3 x 4.

Following is a short summary of five key components of the research on how to build a strong foundation to learn math. Based within the research on how we come to know mathematics and how our mind structures ideas, is our framework for how we build curricular resources and instruct students.

Key component 1: Promoting exploration It is important to help students build new knowledge by connecting it to things they already know. Although this approach might not be intuitive to us as adults, it is important first to allow students time to develop an informal strategy and play with an idea for a short while (which could be 1 to 2 minutes or sometimes longer). Then, by examining these strategies, we encourage students to develop more sophisticated, formal and abstract strategies and algorithms. Our goal, each year, is to help students build a coherent set of schemas to make sense of the mathematics and to use the mathematics to make sense of the world around them. But if they don't recognize when and how to use a more sophisticated strategy, they will always revert to a less efficient strategy or not be able to use any strategy at all.

Key component 2: Enabling modeling   Given a situation or problem to solve (without explicitly having been told how to solve it first), a student needs to determine how to model or represent the problem, and then determine a strategy for how to solve it. Jerome Bruner,[2] a cognitive psychologist, describes how when first learning an idea it is best to model it with a physical model (e.g., cubes), then an iconic or visual model (e.g., number lines), and finally with a symbolic model (e.g., equations or tables). Physically manipulating objects and then diagramming the relationships before moving on to an abstract model helps build a more coherent cognitive structure. These representations provide a student with a base, enabling them to then know when and how to use a more sophisticated abstract model (e.g.,, algorithm).

Key component 3: Strengthening conceptualization   We need to help build a student's conceptual understanding along with procedural understanding. By having a student model the mathematical situation (especially iconically) and then comparing different models with each other, they begin to understand when a method is more efficient in one situation than another. The student is now able to incorporate and organize new information and build a well-connected network of mathematical ideas.

Key component 4: Supporting connection   Knowing the structure of math is critical to building greater knowledge of math concepts. By focusing on structure, students build a deeper understanding of the topics and establish connections among fundamental concepts of math. Structure is defined as elements of the mathematics that remain constant across grade levels. Just as the structure of a building has a foundation, mathematics has common structural ideas that students' learning builds upon over a period of multiple years. For instance, the concepts of unit, composing, decomposing, iteration, partitioning, equivalence and relationships are structural components for number. Understanding that the number 28 is composed of two units of size ten and eight units of size one is necessary to understand place value. Or that by partitioning one into ten

---

[2] Bruner, J. S. (1964). "The course of cognitive growth." *American Psychologist* 19(1): 1–15.

equivalent size units, you get a new unit of one-tenth, which has the relationship of iterating itself ten times to become one. Therefore, maintaining a focus on structure entails helping students see how foundational ideas extend across grade levels and topics. Through gaining an awareness of connections across the different topics, students need not be limited to memorized procedures for each special case and can instead problem-solve in related contexts. Think of the way a skilled auto mechanic would be more adept at fixing a broken lawn mower than a dentist would be. The more you know about the structure of something, physically or theoretically, the more you are able to adapt to unpracticed situations.

Key component 5: Addressing misconceptions It is important to focus on misconceptions, which are ideas that might work in one area of mathematics but do not work in another. By modeling a problem visually, a student begins to understand the parameters of why a strategy might work best in one area and why it is less efficient or doesn't work in another area. For example, after a student learns how to add whole numbers they move to fractions. Many students who add $\frac{2}{3} + \frac{1}{3}$ will add the numerators 2 + 1 = 3 and the denominators 3 + 3 = 6 to get $\frac{3}{6}$. This incorrect answer is not a mistake but a misconception. The vast majority of errors students make in mathematics are the result of deeply held misconceptions that either have been ignored by educators or have been remedied with quick "tricks" and the perpetuation of a focus on procedures as opposed to the root of the student's difficulties. Learning involves navigating mistakes and misconceptions. When we focus on the first four elements discussed above, most mistakes and misconceptions will be minimized, but it is important to acknowledge and address misconceptions. A mistake will often recur even after a student is introduced to a correct procedure, because it stems from deeper mathematical misconceptions. By being aware of why and how misconceptions develop and taking the time to address them through models and discussion, you can move a student to a deeper level of understanding that precludes such mistakes.

MATH FLUENCY

Fluency with basic math facts, particularly multiplication facts, has long been considered a desirable outcome of elementary education. The rationale for the importance of fluency is predominantly the claim that being able to solve simple calculations, or having the facts committed to memory and being able to fluently recall them, will reduce the cognitive load on students' problem-solving processes. Based on the poor performance of fourth grade students on national and international tests of mathematics, it is evident that U.S. students struggle to achieve at a high rate in mathematics. It is possible that a lack of basic fact fluency is part of the impediment to higher achievement as students who can use simple facts as part of larger mathematical activities demonstrate higher math abilities than those students struggling with basic facts. It is precisely this type of finding that implies fact fluency is necessary for high achievement in mathematics and connects fluency to overall success in math. Unfortunately, the impact of the usage of various curricular resources for developing fluency and the relationship between fluency and overall achievement have been only marginally addressed in peer-reviewed publications. The research and historical developments in relation to fluency learning will be addressed in detail in the following section.

Surveying the scope of students' expected learning in late elementary and early middle grades reveals that basic multiplication facts are a prominent subset of mathematical topics as diverse as multi-digit multiplication and division, ratio and proportion, measurement conversions, and fraction concepts. Therefore, it seems reasonable to assign a level of importance to fluency development even if the evidence supporting its value, longitudinally, is relatively meager.

The fluency debate

If you agree with the presumption that fluency is indeed important, it may be surprising to know that math fact fluency is one of the most frequently debated topics in mathematics education. Almost all educators and parents would agree that quick and accurate retrieval of basic math facts is a desirable skill. However, the perceived importance of fact fluency and its emphasis in the classroom varies greatly from class to class. Some educators believe that a drill-and-practice approach, emphasizing memorization and giving students multiple exposures to individual facts, will produce the greatest increase in fact fluency. You may see drill and practice in the use of online math facts practice programs or in the daily use of timed facts tests and flashcards. Others argue that attempting to increase fluency through drill will not enhance students' school experience and that solving word problems and investigating mathematical patterns instead is a more engaging and worthwhile method of building knowledge of facts. This latter perspective is common for teachers who found their own math learning difficult and were unsuccessful when being asked to learn math facts through drill methods in elementary school. Frequently this approach addresses improving students' fluency implicitly, with minimal time spent focusing on how to become faster at recalling basic facts. It is believed that fluency will develop as students are exposed to basic facts in the effort to solve multiple story problems or problems that have multiple steps.

As in many debates in which the most popular positions are binary and polarized, the true answer lies between the extremes. This is certainly the case for fact fluency, a topic that has been studied and debated for almost a century in the United States. We know how to build children's math fact fluency. We've actually known how for decades. The debates and confusion are created by the unfortunate intersection of poor communication between the research community and educators, curricular resources that do not embody what is known about developing fluency, and (to some degree) particular beliefs and orthodoxy regarding what it means to learn math. We discussed the five main reasons for this disconnect in the first section of the book. With regard to fact fluency, neither approach — neither drill nor pure investigations — is good enough for children. Fortunately, examining the history of fact fluency and some of the latest research on the topic can provide a clear answer and lead to an effective, research-based approach to building math fact fluency. That is what this book is all about.

Can drilling math facts be effective?

Some may wonder, "What's wrong with drill?" The reality is that drill is not an inherently *bad* approach to learning some skills. Drill is ineffective in terms of building and retaining fact fluency. This is one of the many instances in which individuals' experiences influence beliefs and therefore trump knowledge. If you are an advocate of drill, think for a moment about what that looks like in a typical classroom. Students likely take timed practice tests each day and then perhaps move on to practicing with flash cards, chanting number facts, or playing drill-based games in pairs or on computers, in an effort to enhance fact fluency. Well then, what happens when there are facts that students don't know? Where, in this drill method, is the instruction or opportunity to improve? If a student doesn't know 7+9 or 6×8 on Monday, what happens to help this student before he or she is subjected to new tests and drills on Tuesday? The theory behind the drill-and-practice approach is actually born from a body of research predominantly conducted on animals and infants, not school children or adults trying to learn something as complex as mathematics. The basic hypothesis was that if students were exposed to facts repeatedly, the information would bond in their minds and be retained. For students with strong numerical memories, this may work for some facts. We all know (or may indeed be) people who seem to have an uncanny ability to remember phone numbers, addresses, or number sequences. But most students do not have this innate memory capacity.

To define effective fluency instruction clearly, it is helpful to look back at changes in education and the long lineage of fact fluency research.

THE RESEARCH AND HISTORY OF FACT FLUENCY

From the 1920s through to the 1960s a group of educational theorists, referred to as **drill theorists**, advocated an approach to teaching that mimicked the popular industrial assembly line. Topics were to be segmented into simple components and students would learn new information by routine drill and practice of these small, isolated skills. The idea was that they would then put these skills together to produce answers to particular math problems. Unfortunately, most students were not successful in either memorizing or applying these facts long-term. Opponents of this approach, frequently referred to as **meaning theorists**, rejected the drill method and advocated an approach that encouraged students to make connections between what they were learning and what they already knew. In classrooms emphasizing this meaning-based approach, students would practice by looking at patterns and finding related facts that could help them solve facts they didn't know. For 7+9, students might know 7+10 and would simply "take 1 away" to derive the sum of 7+9. When solving 6×8, students often know 5×8 and can simply add another 1×8. If you're curious how this method can possibly work, given that it relies on students to already know *some* facts, more details regarding specific instructional strategies to support this flexible approach to fact development will be described throughout this book.

Every ten to fifteen years, beginning around 1925, large-scale studies were conducted comparing students' basic fact fluency in classrooms utilizing either drill or meaning-based approaches. By the late 1960s, all of the largest studies indicated that meaning-based classrooms consistently outperformed drill-based classrooms. To date, no legitimate study has produced significant evidence that drill surpasses the meaning theorists' approach in building fact fluency. There are studies from time to time that appear to imply drill is superior to meaning-based flexible learning of facts; however, they rarely hold up under critical review. There are often many rival explanations as to why drill appeared to be the more effective approach in the study. One common flaw among studies that support drill is **treatment bias**; that is, the groups in the studies are often put under conditions ('treatments', to use an older research term) that favor the drill group.

Studies generally involve testing two groups of students, with one group receiving drill and practice on specific facts, and the other engaging in some alternative mathematical activity. At the end of a pre-determined duration of days or weeks, the two groups are tested again to measure changes in fluency. Frequently, in the studies documenting drill as increasing fluency more than the other group's activities, the test used to measure fluency includes the exact facts that the drill group practiced and also presents the items in a format very similar to that in which the drill group practiced them. The evaluation of what the non-drill group did during the study tends to reveal either minimal detail about their learning experiences or that the non-drill group participated in mathematical activities and content that would not be appropriate to expect fluency development. This would be like testing two basketball players' ability to shoot free throws but having one player shoot from the free throw line during practice sessions and the other player shoot only from the far-left corner of the court. Anyone would expect people practicing a specific skill to outperform those not practicing that skill. This in no way implies that drill is a highly effective approach to teaching basic facts in general. Indeed, most of the research on fluency indicates that flexible thinking and meaningful learning of facts by focusing on mathematical relationships will always be superior to drill and practice.

Why don't schools use research-based fluency development strategies?

So, if drill doesn't work as well as approaches first advocated by the meaning theorists many years ago, why aren't classrooms using a meaning-based approach to teaching fact fluency? Unfortunately, over the years, the fundamental principles underlying the meaning theory have been either watered down or ignored. Evidence pointing to the harmful effects of timed tests on students' self-esteem has often been used to justify never giving timed tests to assess fluency. This does not make sense, as you would need to collect fluency data every 3-4 weeks to know whether students are improving. Various efforts to reform math education in the United States have also struggled to provide a coherent message regarding what teaching fact fluency should look like. The modified, and inaccurate, modern version of the meaning-based approach involves students investigating patterns and solving various word problems. While these methods are crucial to students' conceptual learning of mathematics, they tend to fail in building basic fact fluency because students don't spend enough time actually practicing facts in any form.

Further complicating the study of fluency are the inconsistencies in research-based recommendations for improving students' abilities with basic facts. When studying the method by which students can best become fluent with their facts, authors of such research frequently offer contradictory claims and evidence. Codding et al. (2011)[3] noted in a meta-analysis that, for students who are labeled with learning difficulties or are in the lowest quartile for mathematics, drill and practice (e.g., repetitive and rote memorization) was deemed more effect than other methods. However, additional studies have found that students learning facts by means of flexible and conceptual approaches (e.g., interrelating facts and looking for patterns), increased their fluency rates as well as their ability to transfer these fluent computation skills to non-routine and more complex problems.

In 2015, we conducted a study examining the development of multiplication fact fluency[4]. We compared a drill-based approach to a strategy-based approach. Given the same amount of time practicing for both groups over a four-week period, we found that the strategy-based approach increased multiplication fact fluency to a greater degree than traditional drill and practice techniques requiring repetitious memorization of isolated facts. In the strategy-based approach, students a) anchored difficult facts (e.g., 8×7) to more easily remembered facts (e.g., 8×5) b) visually modeled the related facts with arrays and c) described their methods with peers using specific language scaffolds provided by teachers. We will describe these methods for multiplication fluency development throughout the remainder of this book and explain how similar methods are applicable to the other operations as well.

As it appears that fluency can be increased by disparate methods, the efficiency of the approach used to develop fluency becomes a significant consideration. Our research demonstrated that students using our approach for only ten to fifteen minutes per day improved in fluency over just a five-week time frame. That leaves a significant amount of a school year to maintain or improve fluency with even less frequent practice.

Perhaps more importantly, if one method of fact development has a value beyond just fluency, tangential as it might be, it is logical to suggest that method should be considered superior to the alternatives. Students are not a mere score on a

---

3 Codding, R. S., et al. (2011). "Meta-analysis of mathematic basic-fact fluency interventions: A component analysis." *Learning Disabilities Research & Practice* **26**(1): 36–47.
4 Brendefur, J., et al. (2015). "Developing Multiplication Fact Fluency." *Advances in Social Sciences Research Journal* **2**(8): 142–154.

timed facts test. They come to learning with many different interests, beliefs about themselves as learners, mathematical strengths, and weaknesses. It is reasonable to state, unequivocally, that if we can help young learners become more fluent by means of methods that help them see mathematical relationships, make connections across different math topics, and foster their interest in mathematics, then that is precisely what should be done.

The remainder of this book will outline the exact methods that DMTI has helped educators implement in schools across the United States, to develop students' math fact fluency. We will begin by looking at addition and subtraction, from basic facts to multi-digit fluency.

Fluency Introduction: Addition and Subtraction

What is "fluency" for addition and subtraction? What is "number sense"? And most importantly, how are they developed?

Children develop through a progression as they learn how to add and subtract numbers. Thomas Carpenter, a researcher at the University of Wisconsin — Madison described this progression using four stages: **direct modeling**, **counting**, **derived facts**, and **facts**.[5]

Direct modeling: Children model each part of the problem and do so typically with physical manipulatives such as cubes or chips. For example, when solving 5 + 6 they will count 5 items, count 6 items, push them all together and then count all of them again starting at one and ending at 11.

Counting: Children can "hold"[6] one of the numbers in their head and then count on from the number. To solve 5 + 6 using the counting strategy, you might see the child say "5" and then count on their fingers 6 more to get the answer of 11. You will also see an interesting development in the child's counting skills in that when saying "6" the child will actually only hold up one finger. Try this by putting up your index finger and saying, "6." Then put up the next finger and say, "7." Continue counting on from five until you have six fingers up and say, "11." When children do this, it is an example of double counting—both the total and what has been added

---

[5] Carpenter, T. P., et al. (1999). *Children's mathematics: Cognitively guided instruction*. Portsmouth, NH, Heinemann.

[6] The ability to hold a number in one's head was described by Jean Piaget as quotity and we describe this in the next section.

on to the 5. This process may seem very simple, but it represents a major cognitive development that is vital to be able to use more advanced strategies.

Derived facts: As children develop an awareness of relationships between numbers, they notice that 5 + 5 can help them solve 5 + 6. A derived fact is obtained when a child uses a fact, they know to help them derive a fact they did not already know. Children and adults use this strategy to solve a problem that they do not have the fact for but do it in their heads so quickly (in less than a few seconds) that it appears they know the fact. This strategy is very powerful for building number sense and early algebraic skills, which we will highlight in the next section. This naturally developing strategy is at the core of the DMTI method for building fact fluency. By intentionally practicing these strategies, with learning supports, more children become fluent at a faster rate than if their ability to use derived facts were simply left to chance.

Facts: When children can instantly give the answer, they know a fact. It is generally assumed a student has committed the fact to memory and is simply recalling the information. There is some evidence that some people actually use derived facts strategies so quickly that it appears they are recalling facts from memory.

Unfortunately, too many children get stuck in the counting stage. Adults teach children how to use different strategies to help them get to the derived fact stage, but they still revert to counting on their fingers. It is surprising how many students in later elementary school will count on their fingers for simple addition facts because they never adequately developed their natural abilities to derive sums mentally. For students to be considered fluent with their math facts we need to help them progress to the derived fact and fact stages. But how do we do this? Young children often lack the number sense necessary to see connections between numbers and to use those connections to help them when solving addition and subtraction problems. To help children we need to spend more time focusing on number relationships.

SPATIAL REASONING SKILLS AID IN DEVELOPING FLUENCY

One skill in mathematics that helps children learn facts that are to be remembered and/or used in realistic situations is spatial reasoning. It is one of the cognitive skills most highly correlated with long-term success in mathematics and related subjects. For the purposes of addition and subtraction, spatial reasoning is the ability to recognize how many are in a set not by counting but by observing a visual pattern. For example, knowing by observation which numbers are one/two more or less than any given number. Benchmark numbers of 5 and 10 help children build these spatial relationships, especially when using cubes, bar models, or number lines. By examining part-whole[7] relationships visually, children begin to conceptualize the math and build derived facts and facts that hold over time.

The importance of game play

With lots of experience playing with numbers, children start to recognize number relationships. These are not concepts we can directly teach to a child and expect them to internalize. We cannot tell a child, "4 is one less than 5" and expect them to just memorize this. Board games are an excellent way to build these skills. As one child rolls a 4 while another rolls a 5, the children experience first-hand that 4 moves you one less space than 5. The child begins to internalize and connect those relationships. Numeracy, or number sense, develops from children's experiences with numbers, not direct instruction. So, it is important to provide children with lots of math help through activities, games, language and play.

ADDITION AND SUBTRACTION FLUENCY AND FLEXIBILITY (0-20)

Let us now consider how to build addition and subtraction fluency and flexibility for numbers from 0 to 20. There are four main strategies with modeling and language to help children build and retain these skills: doubles, benchmark, place value and compensation. First, we will look at some informal activities that enable children to start developing fluency from an early age.

When students are first learning addition and subtraction facts, it is important that they have **quotity**, or the ability retain a number in their head without forgetting it. One way to test this with young students is to ask them to count a set of 4-8 objects. Distract them by asking a question such as, What color is your shirt? Then ask them, how many did you just count? If they remember, they have quotity for that size of number. If they are not able to remember, then it is

---

[7] Part-Whole relationships are addition and subtraction problems that examine comparing static situations. For example, "There are 3 blue birds on a limb and 2 yellow birds. How many more blue birds are there than yellow birds?"

important to work on subitizing or 'quickly seeing' activities. (Note: see the next section for subitizing activities.)

Composing and decomposing

One of the first activities to use with children to build their fact knowledge is decomposing and composing numbers. As discussed in the theory section earlier, it is important initially to work with enactive, then iconic, and finally symbolic models. There is great power in being able to visually see the mathematics along with being able to say it and write it out symbolically.

Have children compose and decompose numbers up to 10. Using cubes, have children build a linear model by composing two numbers. Then have them draw a bar model or number line that matches the enacted model, followed by writing an equation or number sentence. If children do not have access to cubes, other objects can suffice in the short term. However, the cubes allow for models that transition more naturally into formal models used in school such as bar models and number lines. As children work with these models and strategies, it is important for them to say the fact out loud once or twice after they complete the activity. For older children, you can try to have them draw the visual model (proportionately) first and then write the equation. Here are some examples with sentence frames to use.

Examples

Compose the number 5 in as many different ways as you can.

Enactive Model	Iconic Model	Symbolic Model (with Sentence Frame)
		4 and 1 compose 5 4 + 1 = 5 1 and 4 compose 5 1 + 4 = 5
		2 and 3 compose 5 2 + 3 = 5 3 and 2 compose 5 3 + 2 = 5

TEMPLATE: Composing a number

Enactive model	Iconic model	Equation (with sentence frame)
		___ and ___ compose ___.
		___ and ___ compose ___.

Ask each child to compose the following numbers: 3, 4, 8, 6, 9, 7, and then 10. They can explore all the combinations as an extension. This process of building and drawing models is an ideal method to help young children learn to write their numbers symbolically. Labeling their drawn models provides a purpose for learning to write the symbols. This reduces the time spent by students repetitiously writing their numbers out, a task that may not be very meaningful for children and could be quite boring.

SUBITIZING

Subitizing is Latin for 'suddenly seeing'. With regard to number, it is important to have the ability to glance at an amount and be able to know it is a number between two and eight. After eight, you would be able to make a good relative estimate of the number under 1000. There are a few different activities you can introduce to a child to improve their subitizing and build quotity. The ability to retain or hold a number in their head is important as a child learns to count on from a base number when adding, or to count back when subtracting.

Activity 1

You can show the child a card with a number of dots on one side (see Template: Dot patterns). Show the child the side with dots for about half a second. Then you ask the child how many dots they saw. If there are four random dots on the card (e.g., card C or G), ask the child whether they just saw all 4, or saw a pattern like 2 and 2 or 3 and 1, or whether they counted all of them starting from 1. If the child responded quickly with the response of 4, then the child has or is building quotity. If there is a slight pause or you observe the child moving their face forward rhythmically then they are probably counting all. Ask them to look at another card and find a pattern on the next card.

TEMPLATE: Dot patterns

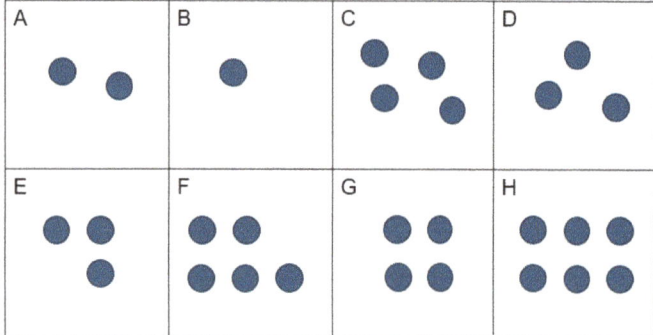

Note: You could use dice and dominoes for this activity, but children tend to memorize the shape and the number name, and not necessarily subitize the number.

Activity 1: Extension   In order to begin building addition and subtraction skills, you can show the child two dot cards and then quickly hide them (e.g., cards C and F). Ask them which number is larger and which number is smaller. Then you ask either, "How many dots would that be all together?" or "How many more dots would we need on the card with the smaller amount to have the same as the larger amount?"

Activity 2

A second activity is to have a handful of cubes (or pennies, small rocks, etc.) — usually up to 6 or 7. Then, hiding them behind your back, put some in your left hand and some in your right. Bringing out one hand, lay the items down in front of the child and then immediately cover them. Ask the child how many they saw. As in the first activity, ask how they saw the number of items and whether they observed any patterns. Looking for patterns builds mathematical thinking. To extend this activity, ask (if they knew the total number of cubes to begin with) how many they think are in the other hand. Then, show these to let them check. You can perform this activity for a few minutes. In the following example, the exchange would go something similar to the following. *How many cubes do you see? I see five. Now I'm going to hide some in my left hand and some in my right hand. How many are in my right hand? Two! How many are still hiding in my left hand? Three!*

  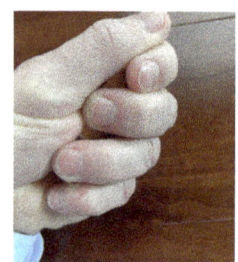

Activity 3

A third activity is to look for objects around their world and try to determine how many of the objects they saw. Here are some different examples. *How many flowers are blooming in the pot (see below)? How many squirrels or birds are in a tree? How many chairs are around the table?*

ADDITION AND SUBTRACTION STRATEGIES

As noted previously, there are four main strategies to help students build fact knowledge, retain basic facts, and solve addition and subtraction problems with and without context: **doubling**, **benchmark**, **place value**, and **compensation**. These represent the four most useful derived facts strategies students should intentionally practice to develop fluency after engaging in the more informal activities described earlier. In this section we explain the mathematics within each strategy, provide examples, and share useful templates.

When choosing strategies and models to use, we always want to consider whether the method has longevity (it can be used across multiple grade levels), it builds flexible thinking, it does not distort place value knowledge, and it encourages early algebraic reasoning. Below we discuss structurally how and why the mathematics works. We use the terms **composing** and **decomposing** with reference to putting together or taking apart the enactive, iconic, and symbolic models. These are the building blocks of the algorithms students will learn later and the different properties such as associative and commutative properties, and identity.

There is another overarching theme to consider before diving into each strategy. When students are learning each of the strategies, there is a progression. Students need to learn each strategy in order with practice to build a strong schema in which to use the strategy and remember the facts years later and in multiple situations. When children are younger and first learning the strategy, it is important to initially enact or build the situation with cubes (or some other physical model), to visually represent it with an iconic model (bar model or number line is best), and finally to represent the situation symbolically (equations). To really reinforce the strategy, the mathematics, and the fact, it is equally important to have the student say the entire process out loud.

Doubling

When numbers are fairly close to each other, you can choose to use a doubling strategy. As an example, let's look at 7 + 8 = ?.[8] Because the numbers are close together, we can choose 7 + 7 to get 14 and then add 1 to get the answer of 15, or we can choose 8 + 8 to get 16 and subtract 1 to get 15. In both cases, we get 7 + 8 = 15.

Structurally, we are composing and decomposing different numbers to find the sum. In the first example we are decomposing the 8 into (7 + 1), in order to compose the two 7s to get 14. Then we compose the 14 with the 1 to get the answer of 15. Mathematically, this is using the associative property: 7 + (7 + 1) = (7 + 7) + 1. However, you can build this strategy and practice it initially without explaining or teaching the mathematical properties. This can come later.

7 + 8 =	
7 + 7 + 1 =	✎ Decompose the 8 into a 7 and a 1
14 + 1 = 15	✎ Compose the 7s
7 + 8 = 15	✎ Restate the fact

Here is what the entire process looks like as you work with one or more students. We will continue to use 7 + 8 as an example.

Enactive model

Using large construction paper, ask the students to use cubes[9] and build a 7 and an 8. Next, decompose the 8 into a 7 and a 1. Have them observe and point to the two sevens. They can even line them up to check they are the same length. Then, ask them to connect them together. Ask how much they have altogether. Using sentence frames such as, "I know ___ and ___ compose/make ___." "I know 7 and 7

---

[8] The numbers being added together are called **addends** because in Latin 'add' means to increase. The answer is called the **sum** or total.

[9] It is best to use cubes (or blocks or snap cubes) to create a linear model. This has an additional benefit of building relative quantity and proportion — two foundational ideas embedded throughout mathematics. You can use chips or tiles and line them up as well, but these tools are not as effective over time.

compose 14." Now, have the student add the last 1 to the 14 and then ask them to use the sentence frame again. "I know 14 and 1 compose 15." Finally, ask the student to repeat the initial fact. "So, I know 7 + 8 = 15." It is important to note that students may not know the sum of 7 + 7 initially. But, by having the cubes available the student can find that sum, and by using the double as an anchor fact to connect to 7 + 8, they are actually more likely to remember 7 + 7 in the future. This is why the language scaffolds we have provided encourage students to state, "I know 7 + 7 compose 14." That statement will instill a sense of confidence and serve as fact practice for the double.

Iconic model

Use a bar model (to match the blocks) or a number line to visually represent the doubles strategy. In the past, this representation had often been skipped to get right to the symbolic representation. However, the ability to spatially draw and see the mathematics is highly correlated with later success in higher levels of mathematics (e.g., proportional reasoning and graphing functions).

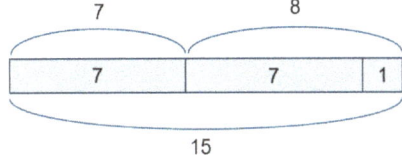

Symbolic model

Ask them to use expressions and equations to match the process they went through with the enactive and/or iconic model for building the doubles strategy.

    7 + 8 =

    7 + 7 + 1 =

    14 + 1 = 15

    7 + 8 = 15

During, or at least at the end, have the student use a sentence frame. Here is an example: "I know ___ and ___ compose/make ___." "I know 7 and 7 compose 14." Now, have the student add the last 1 to the 14 and then ask them to use the sentence frame again. "I know 14 and 1 compose 15." Finally, ask the student to repeat the initial fact. "So, I know 7 + 8 = 15."

Doubles strategy: examples

Expression	Enactive model	Iconic model
7 + 8 =	(enactive model image: bars showing 7 + 7 + 1 = 15)	(iconic model: 7 and 8, with 7 + 7 + 1 = 15)
	Symbolic Model	**Sentence Frame**
	7 + 8 = 7 + 7 = 14 14 + 1 = 15 7 + 8 = 15	I know 7 + 7 is 14. I know 14 + 1 = 15. So, 7 + 8 = 15
	Enactive Model	**Iconic Model**
7 + 8 =	(enactive model image: bars showing 8 + 8 − 1 = 15)	(iconic model: 8 and 8, with 1 crossed out, 7 and 8, total 15)
	Symbolic Model	**Sentence Frame**
	8 + 8 = 16 16 − 1 = 15 7 + 8 = 15	I know 8 + 8 = 16. I know 16 − 1 = 15. So, 7 + 8 = 15

We tend not to use a doubles strategy for subtracting, but it can be used for single digit subtracting. Let's try an example with 9 − 7 = ? .[10] In the case of subtracting, we use the **subtrahend** or the amount we are taking away for the double. We would decompose the first number to leave the amount that matches the subtrahend. I know 9 is 2 + 7. And I know 7 − 7 = 0. So, 9 − 7 = 2. Or, in the case of a fact such as 9 − 4 it is possible to think of a related addition double such as 4 + 4 = 8 meaning 9 − 4 = 5 because 9 is one more than 8. We will examine this strategy in more detail

---

10 The first number in subtraction is called the **minuend** because the Latin stem 'minu' means to decrease or diminish. So, the minuend is the number that will be decreasing. The second number is the **subtrahend**. The Latin stem 'subtra' means to take away, so that number is what is being taken away. The answer is referred to as the **difference**.

in the multidigit section as the larger numbers make for even more interesting doubles combinations.

Benchmark strategy

The benchmark strategy is one of the most powerful and flexible ways of adding and subtracting numbers. There are two key cognitive elements that students must have to be able to use this strategy efficiently and over time. The first element is the ability to find the next benchmark number, which is typically a ten or five depending on the size of the number. The second element is the ability to decompose a number in relation to the benchmark number. Here is an example.

If we are adding 7 + 8, then we can choose either the 7 or the 8 and state that the benchmark number is 10. If we choose 7, then we would determine that 7 is 3 away from 10. This means that we need to decompose the 8 into a 3 and a 5. So, 7 + 3 = 10 and then 10 + 5 = 15. Therefore, 7 + 8 = 15. This specific use of the benchmark strategy is prominently referred to as 'making 10' in educational research.

As with other strategies, we use the structural words of decomposing and composing. Mathematically, we are using the associative property.

7 + 8 =	✎ Determine 7 is 3 away from the benchmark number 10
7 + (3 + 5) =	✎ Decompose the 8 into 3 and 5
(7 + 3) + 5 =	✎ Use the associative property to compose 7 and 3
10 + 5 = 15	✎ Compose the 10 and 5 to get 15
7 + 8 = 15	✎ Restate the fact

The benchmark strategy can also be used for smaller numbers. For example, if you are adding 4 + 3, a student might say, "I know four plus one is five. I took one from the three. So, five and two more is seven. So, four plus three is seven."

It also works for some double digits and single digits. We will use 27 + 8. A student (instead of adding the ones digits) might see 27 + 3 = 30. So, they can decompose the 8 into a 3 and 5. Next they would add 30 and 5 to compose 35. So, 27 + 8 = 35.

The benchmark strategy also works well with subtraction problems. Young children can sometimes need more time using the benchmark strategy for addition than they do to use doubles effectively. However, it is precisely the value the benchmark strategy offers for subtraction fluency that justifies the ample practice time. Subtraction benchmark strategies work similarly to the addition approaches, but instead of finding the next benchmark number that is larger, we now look for the next smaller benchmark number. For example, if we are given 13 − 8 = ?, then we would say 13 is 3 away from 10. So, we would decompose 8 into a 3 and a 5. We first take 3 away from 13 to get 10 and then take 5 away from 10 to get 5. So, 13 − 8 = 5. Having students relate their benchmark strategy for 13 − 8 to 5 + 8 is also a powerful connection that helps students use benchmarks. They begin to recognize they are using the same benchmark number and composing and decomposing in the same way. The only difference is whether they are adding up or counting back.

13 − 8 =

13 − 3 = 10

10 − 5 = 5

13 − 8 = 5

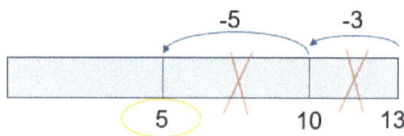

We have been working with subtraction as 'take-away'. However, subtraction can also be viewed as the difference or 'distance' between two numbers. This view allows for the connection to addition described earlier to be even clearer to children. For 13 − 8, we would be asking what is the distance, or difference, between 8 and 13. In this case, we would determine the distance to the next benchmark number in either direction. Starting at 8, we would say 8 is 2 away from 10 and 10 is 3 away from 13. And, 2 and 3 is 5. So, 13 − 8 = 5. Or 10 is 3 away from 13 and 8 is 2 away from 10. Because 3 and 2 is 5, 13 − 8 = 5.

13 − 8 =

8 + 2 = 10

10 + 3 = 13

2 + 3 = 5

13 − 8 = 5

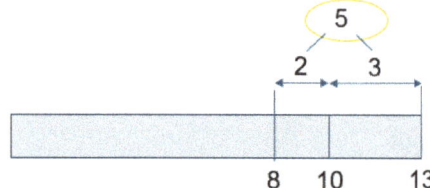

Here are some examples of how to model the benchmark strategy and how to use structured language to explain the process. Initially, we will use smaller numbers with the enactive models and then move to iconic and symbolic models with sentence frames. We will use the example 6 + 9.

Enactive

Have students use cubes to build 6 + 9. Then, ask them, "Starting from 6, what is the next benchmark number?" "10." "How many more from 6 to make 10?" "4." "So, what do we do next?" Have the student use the word decompose. "I need 4 from the 9. So, I decompose 9 into a 4 and a 5." "That makes 10 and 5, which is 15." "So, 6 + 9 = 15."

Iconic

If the student built the enactive model with cubes, then have the student draw a bar model that matches how they decomposed the numbers. Or you could ask them to first use a bar model to show the benchmark strategy. Ask them to explain out loud all of the steps they took. For decomposing the 9, their bar model should be like the one below and they should say something like the following. "I'm using the benchmark strategy and 6 is close to 10. So, I decomposed the 9 into a 4 and a 5. I did this to compose 6 and 4 to make 10. That leaves 10 and 5, which is 15. So, 6 + 9 equals 15."

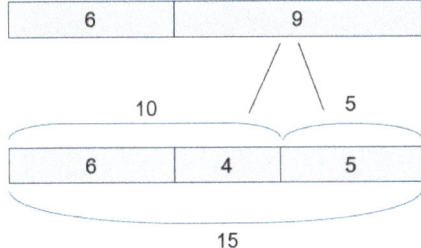

Then, ask the student if they could use the benchmark strategy in another way for this problem. This helps them become more flexible in their thinking and to look for relationships. You can ask them to create another bar model to match their thinking and end by having them explain the strategy, which would be similar to the following. "I know 9 is 1 away from 10. So, I can use the benchmark strategy and decompose 6 into a 5 and a 1. I gave 1 to 9 to make 10 and that left 10 plus 5, which is 15. So, 6 + 9 is 15."

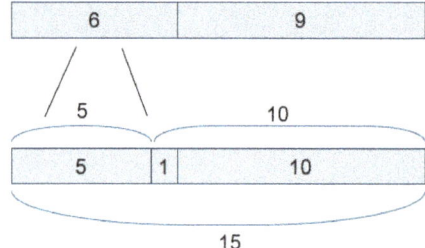

Symbolic

Have the students write down each of the steps for the benchmark strategy. Encourage them to try both numbers and decide which is easier and why.

6 + 9 =	
6 + 4 + 5 =	I decomposed 9 into 4 and 5.
10 + 5 =	I composed 6 and 4 to make 10.
10 + 5 = 15	I composed 10 and 5 to make 15.
6 + 9 = 15	So, 6 + 9 = 15.
6 + 9 =	
5 + 1 + 9 =	I decomposed 6 into 5 and 1.
5 + 10 =	I composed 1 and 9 to make 10.
10 + 5 = 15	I composed 10 and 5 to make 15.
6 + 9 = 15	So, 6 + 9 = 15.

"I think decomposing the 6 was easier, because then I just had to add 1."

MORE ADVANCED STRATEGIES

These next two strategies are slightly more complex than the use of doubles and benchmarks. However, they are invaluable for children to practice meaningfully because they allow for work with larger numbers and connect to topics children will learn in later grades (e.g., algorithms and algebraic expressions).

Place value strategy

A place value strategy has to include at least one two-digit number, otherwise you would use one of the other strategies. In whole numbers (no decimals or fractions), each place to the left is ten times the value. For example, the number 32 has two digits. The digit 2 is in the ones place and each unit is worth 1, making a total of 2. The digit 3 is in the tens place and each unit is now worth 10, making a total of 30. To use the place value strategy, we decompose each of the two-digit numbers by its place value. Then, we compose equivalent units across the two numbers: ones with ones and tens with tens. Mathematically it does not matter, but typically it is easier to put together larger units and then smaller units.

Before using this strategy, students need to understand how each digit represents its value using powers of a ten. If students are struggling, this means they need more practice on place value. One activity would be for students to draw a number line (or scaffold this activity with a given number line) with no numbers. Have the student place where 4 is. Then add 10. Ask the student to find and place the number on the number line. It should be 14. Ask the student what stayed the same and what changed. The student should say the ones digit is still 4, but the tens digit changed from 0, or nothing, to 1. Ask, "Why?" The student should say, "Because I did not add any units of one, but I did add a unit of ten." The use of the term unit is critical in that it clarifies that the digit in each place is a count of those particular units. Add 3 to the last number, then 20, then 2 and then 10. Each time the same set of questions should be asked regarding the digits staying the same and changing.

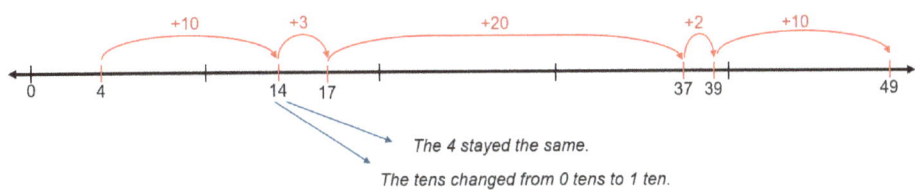

The 4 stayed the same.
The tens changed from 0 tens to 1 ten.

Once students have a grasp of place value, then they can use the place value strategy to add and subtract numbers. Here is the mathematics behind the strategy for 12 + 13. First, we decompose each of the numbers by their place values to get 10 + 2 and 10 + 3. Next, we compose similar units using the associative property: (10 + 2) + (10 + 3) = (10 + 10) + (2 + 3). Then we add tens (10 + 10 = 20) and add the ones (2 + 3 = 5). Finally, we compose the tens and ones together (20 + 5 = 25) and end by referring back to the original statement (12 + 13 = 25). In later years, students will learn to refer to this strategy as **partial sums**, because you find the sum in parts. Partial sums is actually the method most useful for teaching students the standard addition and subtraction regrouping algorithms. Therefore, using the place value strategy for fluency development is supporting students' long-term success in mathematics. Below is an example of what this process looks like when presenting it to students through the different representations.

Enactive model

When building the place value strategy, either have the students use Cuisenaire rods that have separate tens and ones units or use painters' tape to wrap sets of 10 cubes together, allowing the students to see that they now represent a new unit (that what used to be 10 units of 1 have now become 1 unit of 10 — a critical understanding). Have the student build a linear model of 12 + 13 by laying out a unit of 10 and 2 ones and then another unit of 10 followed by 3 ones. It is more powerful for students to see the horizontal linear model (that mimics the number line) than just having the tens and ones in a vertical model. Then have the students rearrange the units so the tens come first and then the ones. Ask, "How many units of 10 are there together?" The student will respond, "2." Ask, "How much is that worth altogether?" "20." Then ask, "How many units of 1 are there altogether?" "5." Now ask, "How much is it altogether?" "25." And, finally, ask the student to rephrase the original problem. "So, that means 12 + 13 = 25."

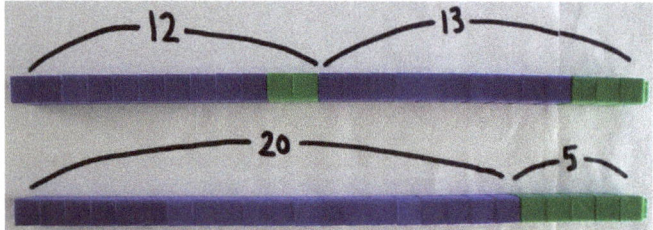

Iconic model

The iconic model could be either a number line or a bar model. We are going to represent the process with a bar model to represent a clearer match between the enactive and iconic models. The questions should match the above process. (Please note that it does take students some time to draw these models. However, drawing the model strengthens their conceptual knowledge of the mathematics, including place value, proportion, addition, and the identity property.

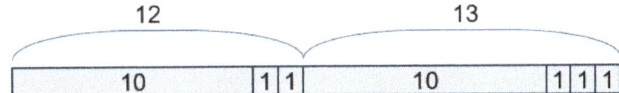

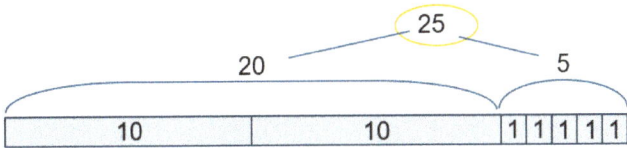

Symbolic model

For the symbolic model, the students can write the equations in one line, as long as the equal sign is used correctly. It would look like this: 12 + 13 = 10 + 2 + 10 + 3 = 10 + 10 + 2 + 3 = 20 + 5 = 25.[11] We prefer students to write the equations vertically, so you can ask questions.

12 + 13 =

10 + 10 + 2 + 3 =   I can decompose 12 into a 10 and a 2, and 13 into a 10 and a 3.

20 + 5 = 25   I compose the 2 tens to get 20, and the 2 and 3 to get 5.

12 + 13 = 25   So, 12 plus 13 is 25.

Figure. Place value strategy example for addition

Expression	Iconic model	Symbolic model	Sentences
12 + 6 =		10 + 2 + 6 = 10 + 8 = 18 12 + 6 = 18	I can decompose 12 into a 10 and a 2. 2 and 6 compose 8. 10 and 8 compose 18. So, 12 plus 6 is 18.

---

11 A common but incorrect use of the equal sign looks like this: 12 + 13 = 10 + 10 = 20 + 2 + 3 = 5 = 25.

TEMPLATE: Place value strategy

Use the place value strategy to solve the following:

Expression	Iconic model	Symbolic model	Sentences
13 + 5			
14 + 13			
9 + 16			
23 + 16			

Place value strategy for subtraction

The place value strategy also works for subtraction problems. However, there are some possible misconceptions that need to be addressed. Here is the first example: 15 – 12. For subtraction, the first number (minuend) is the amount we have, and the second number (subtrahend) is how much we are going to take away. So, we can decompose the first number to 10 and 5. Then, we can decompose the second number into 10 and 2 but must remember we need to take away 10 and then take away another 2. Structurally, we are still using the associative and distributive properties: 15 – 12 = (10 + 5) – (10 + 2) = 10 – 10 + 5 – 2 = 3. The subtraction is using the distributive property: – 10 and – 2.

However, for younger students, the following notation and language is more acceptable. Let's try 15 – 8 with the place value strategy. First, we decompose by place value and get 10 + 5 – 8. Because there are no tens to subtract, we move to the ones. We ask the student, "Do we have enough ones to take away?" The response should be, "I only have 5 ones and I need to take away 8."[12] Because we do not have enough ones then we can decompose the 8 into 5 and 3. This allows us to take away 5 ones. The final step is to take away the remaining 3 ones from the 10 to get 7. Structurally, it is looks like this: 15 – 8 = 10 + 5 – 8 = 10 + 5 – 5 – 3 = 10 – 3 = 7. This strategy has a number of steps so it should not be taught to students until they have a grasp of subtraction and decomposing.

---

[12] Note: many teachers and parents (from misguided curricular resources) will teach students to say "I can't subtract 8 from 5." Please discourage this statement. Yes, we can take 8 away from 5 and it is –3. Algebraically, we do this all the time and there is another algorithm that actually uses this method.

Figure. Place value strategy examples for subtraction

Expression	Iconic model	Symbolic model	Sentences
15 – 12 =		15 – 12 = 10 + 5 – 10 – 2 10 – 10 = 0 5 – 2 = 3 15 – 12 = 3	I can decompose 15 into a 10 and a 5. I can decompose 12 into take away 10 and take away 2. 10 take away 10 is 0 and 5 take away 2 is 3. So, 15 take away 12 is 3.
15 – 8 =		15 – 8 = 10 + 5 – 5 – 3 5 – 5 = 0 10 – 3 = 7 15 – 8 = 7	I can decompose 15 into a 10 and a 5. I can decompose 8 into take away 5 and take away 3. 5 take away 5 is 0. Then, 10 take away 3 is 7. So, 15 take away 8 is 7.

Compensation strategy

We typically introduce compensation as the last of the addition and subtraction strategies. It is not necessarily the most difficult of the strategies (many students actually prefer to use compensation over some other strategies), but mathematically it is more difficult to explain and it is only practical for particular number combinations. You can introduce the strategy, but then can decide when to have students explain how it works.

When one or both numbers are close to a benchmark number then this strategy is very useful. Here is how it works for 7 + 9. Nine is close to ten, so we could add 7 + 10 = 17. But we did not have 10, we had 9, which is one less. So, we need to subtract one from 17 to get 16. This means that 7 + 9 = 16.

Here is what is happening structurally. As with the benchmark strategy, we are using a benchmark number. However, with compensation we are not decomposing the other number to compose a benchmark number. In this case, we are using two important properties in mathematics. First, we use the **identity property of addition**, which states that when you add zero to a number then that number does not change. But then we use the **additive inverse property** that states that when you add a number to its opposite (the directed number with the opposite sign), you get zero (e.g, 3 − 3 = 0 or 2 − 2 = 0). This is a technique used in algebra and other upper-level mathematics courses. Mathematically it looks like this:

7 + 9 =

7 + 9 + 0 = identity property where adding 0 does not change the sum

7 + 9 + (1 − 1) = ? Additive inverse to subtract a number from itself to get 0

7 + (9 + 1) − 1 = ? Use associative property

(7 + 10) − 1 = ? Add the two addends

17 − 1 = 16 ? Subtract the remaining amount

7 + 9 = 16 ? Restate the facts

Figure. Compensation strategy examples

Expression	Enactive model	Iconic model
7 + 9 =		
	Symbolic model	**Sentence frame**
	7 + 10 = 17 17 − 1 = 16 7 + 9 = 16	I know 7 and 10 is 17. I know 17 take away 1 is 16. So, 7 + 9 = 16.
	Enactive model	**Iconic model**
8 + 18 =		
	Symbolic model	**Sentence frame**
	8 + 20 = 28 28 − 2 = 26 8 + 18 = 26	I know 8 composed with 20 is 28. I know 28 take away 2 is 26. So, 8 plus 18 is 26.

A compensation strategy can be used with subtraction as well. Allow students to practice compensation with addition for a while before you introduce compensation with subtraction. Students need to be able to attend to what they are compensating and remember to do the opposite or inverse operation. We will go through a couple examples.

If we are subtracting 18 – 13, we could compensate the first number (minuend) to 20. So, 20 – 13 = 7. We compensated by adding 2 to the 18 to compose 20, so we need to subtract 2 to get the final difference. 7 – 2 = 5. That means 18 – 13 = 5. Remember, structurally we actually added 0 to 18 using the identity property (2 – 2).

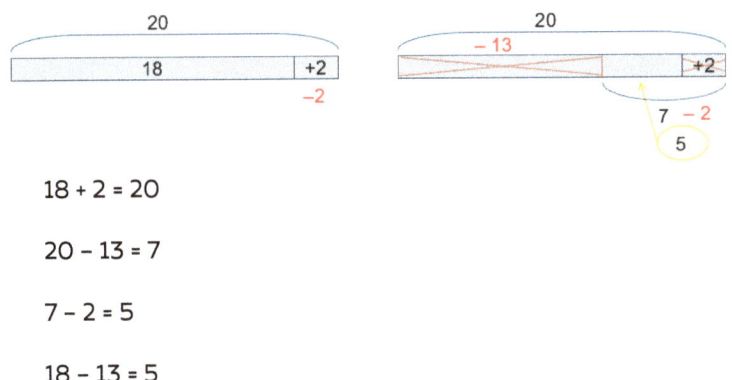

18 + 2 = 20

20 – 13 = 7

7 – 2 = 5

18 – 13 = 5

For 14 – 9, we might want to compensate the 9. I know 14 – 10 = 4. The key is to ask students to focus on what the operation is. We were first subtracting nine and now we are subtracting ten. This means that we subtracted one too many, so, to compensate, we need to add one back, so 4 + 1 = 5. This means 14 – 9 = 5. Note that many students will want to say, "We added one more to 9 to make 10, so we need to subtract one." These students will incorrectly get three. The misconception is forgetting that the 10 itself is being taken away not added.

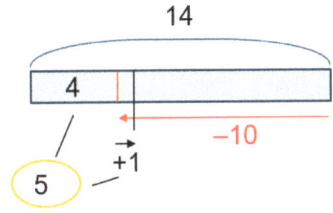

14 – 9

9 + 1 = 10 (but we are taking away 10)

14 – 10 = 4

4 + 1 = 5

14 – 9 = 5

ADDITION AND SUBTRACTION STRATEGY SUMMARY

This chapter focused on introducing the four main strategies to practice and build addition and subtraction fluency through improving students' flexibility. For younger students, it is important to introduce each strategy by itself and practice it by varying the number sets and having students model it using enactive, iconic, and symbolic representations. Students should always practice saying the steps out loud and coming back to the original problem they are solving. A rule-of-thumb would be to let them practice each strategy for a month with supervision and scaffolding before combining them. These are typically introduced in schools in first and second grade. By third grade and up, students should be able to use all four strategies. It is important to have them practice these different strategies each year.

Use the template at the end of the chapter to add your own problems. You can print it on larger paper and then have students complete each of the strategies and have them complete and return the template in a week or two. Once they have completed it correctly, have students pair up and roll two dice. The first die is the problem and the second die is the strategy. If you roll a 5 or 6 the student gets to choose any addition or subtraction problem. Then the other student has to verbally solve the problem by stating the strategy. If a strategy doesn't work particularly well for a given fact, have students explain why the strategy is not the most useful for that specific fact. For example, a doubles strategy is less helpful for 9 + 5 than a benchmark or compensation strategy would be.

There is a progression of building the models. Initially, the enactive model should be used with an iconic and symbolic model. Then, students should model the problems iconically and symbolically. Finally, they can model problems with equations. Again, each time they should state the process verbally. If students are struggling with the equations or verbally explaining the process, it is important to reverse the process and iconically diagram the problem or even enactively build the model with cubes. Remember, people who are successful with mathematics have the ability to be flexible with their strategies, have a visual model in their brain, and articulate each process.

Fluency Introduction: Addition and Subtraction

TEMPLATE: Addition and subtraction strategy practice (0–20)

Expression	Iconic model	Symbolic model	Sentences
6 + 8			
7 + 9			
12 + 13			
18 + 15			

Expression	Iconic model	Symbolic model	Sentences

Addition and Subtraction Fluency and Flexibility with All Numbers

Fact fluency develops when students become more flexible with numbers and have more strategies. It is critical to be fluent and flexible and this is built through a focus on three components: a) working on the four addition and subtraction **strategies**, which builds familiarity with number relationships, b) varied **practice** through enactive, iconic, and symbolic models, and c) **language** scaffolds.

Keep in mind that it is important for students to be able to visualize the mathematics, be able to compute correctly, and be able to explain the mathematics. These work together to build a strong foundation for knowing and remembering math content. This developmental process is analogous to becoming a good long-distance runner. It is important to run, but not to run every day. To become a better, faster runner, you need to interperse running with lifting weights, stretching and doing some high-intensity workouts.

FLUENCY AND FLEXIBILITY FOR MULTI-DIGIT NUMBERS

Once students have worked on and practiced the four strategies (doubles, place value, benchmark, and compensation) with numbers under twenty, they can begin working on other number sets. Have students work on multi-digit problems and encourage them to use and name the processes; they will improve both their fluency and their flexibility with adding and subtracting numbers. You do not need to have them work on all the strategies each time. The practice now can, and should, vary. We will not be highlighting the enactive model, but if students are struggling with these number sets, you should a) have them model the problem with physical objects, and/or b) modify the number sets to make them easier. Students also will not need to always create an iconic model, but should do so from time to time, in order to continue building a foundation.

Let's try an example: 17 + 19. Initially, ask the student to create an iconic model (number line) and a symbolic model (equations) for each of the four strategies. Below, we provide some examples on how best to get them to explain their processes.

Doubles

First, students should ask themselves whether the numbers are close and whether they know a suitable doubles fact. If they do not know the doubles using one of the addends (e.g., 17 + 17 or 19 + 19) then they could find another one they know (e.g., 15 + 15). Here are the two most typical.

Iconic model

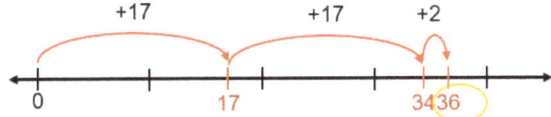

Symbolic model

17 + 19 = 17 + 17 + 2	I decomposed the 19 into a 17 and a 2.
17 + 17 = 34	I know the doubles fact that 17 and 17 is 34.
34 + 2 = 36	I compose the 2 to 34 to get 36.
17 + 19 = 36	So, 17 plus 19 is 36.

If the student does not know this doubles fact, then ask if they know another one.

17 + 19 = 15 + 2 + 15 + 4	I decomposed the 17 into 15 + 2 and the 19 into 15 + 4.
15 + 15 = 30	I composed the doubles fact 15 and 15 is 30.
2 + 4 = 6	I composed the 2 and the 4 to get 6.
30 + 6 = 36	I composed the 30 and the 6 to get 36.
17 + 19 = 36	So, 17 plus 19 is 36.

As the numbers in the problems change, the usefulness of doubling with multi-digit numbers can vary. There will likely be many problems in which it will take some careful thought to decide what double might even be useable. It is important for students to explore using doubles, even in instances in which they feel another strategy is more prudent or in which a useful double is not readily apparent. This exploration will increase number sense, place value knowledge, and accuracy with any strategy the students ultimately use.

Benchmark strategy

Using the benchmark strategy for 17 + 19, the student should first state which number they are going to use to find the next benchmark number and which number they are going to decompose. They might say the following, "I'm going to use 17. So, the next benchmark number is 20. I'll decompose the 19 into a 3 (because I need 3 to add to 17 to make 20) and a 16.

Iconic model

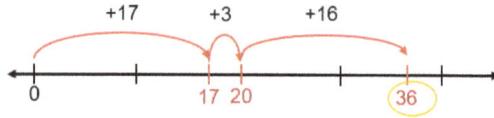

Symbolic model

17 + 19	
17 + 19 = 17 + 3 + 16	I decomposed the 19 into a 3 and a 16.
17 + 3 = 20	I compose 17 and 3 to make a 20 (the benchmark number).
20 + 16 = 36	I composed the 20 and 16 to make 36.
17 + 19 = 36	That means 17 plus 19 is 36.

Place value strategy

For the place value strategy, students will decompose each number by place value and then compose those numbers to get the remaining sum.

Iconic model

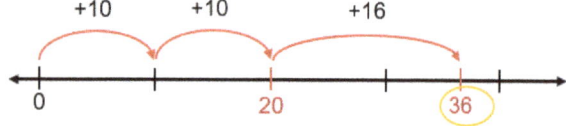

Symbolic model

17 + 19 = 10 + 7 + 10 + 9	I decomposed each number by place value.
10 + 10 = 20	I composed the tens to get 20.
7 + 9 = 16	I composed 7 plus 9 to get 16.[13]
20 + 16 = 36	I composed 16 and 20 to get 36.
17 + 19 = 36	So, 17 plus 19 is 36.

Compensation strategy

Although the compensation model looks similar to the benchmark strategy, it is different in that you are adding a new number to get to the benchmark number and then must compensate for that number to get the answer. Mathematically, you are using the additive identity and the additive inverse properties. In this example, you would compose 3 to 17 to make 20. In the equation, you are actually adding 0, the additive identity property to not change original sum, but you are doing so in the form of (+3 − 3), which is the additive inverse. So, after you add the 19 to 20 to get 39, you must compensate and subtract 3 to get the answer of 36. The examples given are situations in which compensation is helpful. There may be number combinations in problems students are working with that do not necessarily lend themselves to compensating. For example, 23 + 29 is an ideal time compensate and use 23 + 30 − 1, but 23 + 24 is a problem in which compensation may not seem particularly useful. Encourage students to explore how they could use compensation even when it is not ideal and have them articulate why another strategy is more sensible. They should ultimately choose their preferred strategies but practicing unfamiliar or more difficult approaches can improve students' future choice of strategies and computational accuracy.

Iconic model

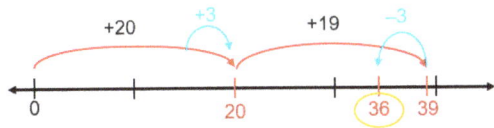

---

13 Ask the student whether they just know 7 + 9 = 16. Encourage them to use a strategy if they are counting on from ones. I can decompose 7 into a 6 and 1. 1 plus 9 is 10 and 10 and 6 is 16. So, 7 + 9 = 16.

Symbolic model

17 + 3 + 19 − 3	I want to add 3, so I know I must subtract 3 later.
17 + 3 = 20	I composed 3 to 17 to get 20.
20 + 19 = 39	I composed 20 and 19 to get 39.
39 − 3 = 36	I compensated and subtracted 3 from 39 to get 36.
17 + 19 = 36	So, 17 plus 19 is 36.

Doubles and compensation strategy

Sometimes a student might use two strategies together. Here is an example where a student might notice they could use a doubles strategy for adding 19 to get 38. Then they have to compensate the two they added to 17 to get the other 19.

17 + 19 = 17 + 2 − 2 + 19	I use a compensation strategy of adding 2 to 17 to get 19.
19 + 19 = 38	I know the doubles fact 19 plus 19 is 38.
38 − 2 = 36	I use the compensation and subtract 2 from 38 to get 36.
17 + 19 = 36	So, 17 plus 19 is 36.

FLUENCY AND FLEXIBILITY WITH DECIMALS AND FRACTIONS

Now that students have worked with whole numbers, and if they are ready, they can practice with decimals and fractions. The idea here is not to use these strategies to solve all problems involving addition and subtraction of decimals and fractions. However, these strategies can be used to build fluency and flexibility with decimals and fractions. By using these techniques to improve students' number sense, you will be more likely to see a reduction in the computational errors that often plague students when they first learn to operate with decimals and fractions. Additionally, intuitive estimation skills will be enhanced, which also improves accuracy.

With decimals, we suggest you focus on decimal numbers between 0 and 5. We will use tenths in our examples. These items really help students at the intermediate levels understand place value as well. With fractions, it is important to work with sets such as halves, fourths and eighths; fifths and tenths; or thirds, sixths and ninths.

We will use 1.6 + 1.8 to highlight each strategy for decimals.

1.6 + 1.8	Bar model (iconic)	Equations (symbolic)	Sentences
Doubles	[bar model: 1.5 + 1.5 = 3, then 0.1 + 0.3 = 0.4, total 3.4]	1.5 + 1.5 = 3.0 0.1 + 0.3 = 0.4 3 + 0.4 = 3.4 1.6 + 1.8 = 3.4	I doubled 1.5 to get 3.0. I composed 0.1 and 0.3 to get 0.4. Then I composed 3 and 0.4 to get 3.4. So, 1.6 and 1.8 is 3.4.
Place Value	[bar model: 1, 1, 0.8, 0.6; 2 and 1.4; total 3.4]	1 + 1 = 2 0.6 + 0.8 = 1.4 2 + 1 = 3 3 + 0.4 = 3.4 1.6 + 1.8 = 3.4	I composed 1 and 1 to get 2. I composed 6 tenths and 8 tenths to get 14 tenths or 1.4. I compoosed 2 and 1 to get 3. Then I composed 3 and .4 to get 3.4. So, 1.6 and 1.8 is 3.4.
Benchmark	[bar model: 1.6, 0.4, 1.4; 2 and 1.4; total 3.4]	1.6 + 0.4 = 2 2 + 1.4 = 3.4 1.6 + 1.8. = 3.4	I composed 1.6 with 0.4 to get 2. By decomposing the 1.8 I get 0.4 and 1.4. I composed the 1.4 with 2 to get 3.4. So, 1.6 and 1.8 is 3.4.
Compensation	[bar model: 1.6, 2; 3.6; −0.2; total 3.4]	1.6 + 2 = 3.6 3.6 − 0.2 = 3.4 1.6 + 1.8 = 3.4	I added 0.2 to 1.8 to get 2.0 I composed 1.6 and 2 to get 3.6. I comensated the 0.2 and subtracted it from 3.6 to get 3.4. So, 1.6 and 1.8 is 3.4.

We will use $1\frac{3}{4} + 2\frac{1}{8}$ to highlight each strategy for fractions. When working to build students' fluency and flexibility with fractions, it is imperative to ensure that they first understand fractions conceptually and are adept at using models and structural language to represent them. One key idea that must be in place before students begin to try to improve their fluency with fractions is the concept of a **unit fraction**.

MATH FACTS

Unit fractions are any fraction with a numerator of 1 and represent the unit the fraction is using to measure the total quantity. The fraction $\frac{3}{4}$, it is accurately described as "three one-fourth units." They can be written in mathematical notation as $\frac{3}{4} = 3 \left(\frac{1}{4} \text{ units}\right)$.

In the example below, the fractions being added offer easy conversions to a common denominator. Fluency with fractions is most effectively developed when the fractional units involved are fairly common and students are able to complete simple conversions and estimates with them. Halves and fourths, fourths and eighths, thirds and sixths, or fifths and tenths are all combinations likely to be familiar to students and will work better for building confidence than trying to have them operate fluently with, for example, thirds and sevenths.

$1\frac{3}{4} + 2\frac{1}{8}$	Bar model (iconic)	Equations (symbolic)	Sentences
Doubles		$1\frac{3}{4} + 1\frac{3}{4} = 3\frac{1}{2}$ $3\frac{4}{8} + \frac{3}{8} = 3\frac{7}{8}$ $1\frac{3}{4} + 2\frac{1}{8} = 3\frac{7}{8}$	I doubled $1\frac{3}{4}$ to get $3\frac{1}{2}$. I know $\frac{1}{2}$ is $\frac{4}{8}$ and I composed $3\frac{4}{8}$ and $\frac{3}{8}$ to get $3\frac{7}{8}$. So, $1\frac{3}{4}$ and $2\frac{1}{8}$ is $3\frac{7}{8}$.
Place Value		$1 + 2 = 3$ $\frac{3}{4} + \frac{1}{8} = \frac{7}{8}$ $3 + \frac{7}{8} = 3\frac{7}{8}$ $1\frac{3}{4} + 2\frac{1}{8} = 3\frac{7}{8}$	I composed 1 and 2 to get 3. I composed $\frac{3}{4}$ and $\frac{1}{8}$ to get $\frac{7}{8}$. I composed 3 and $\frac{7}{8}$ to get $3\frac{7}{8}$. So, $1\frac{3}{4}$ and $2\frac{1}{8}$ is $3\frac{7}{8}$.
Benchmark		$1\frac{3}{4} + \frac{1}{4} = 2$ $2 + 1\frac{7}{8} = 3\frac{7}{8}$ $1\frac{3}{4} + 2\frac{1}{8} = 3\frac{7}{8}$	I composed $1\frac{3}{4}$ and $\frac{1}{4}$ to get 2. I know $\frac{1}{4}$ is $2\frac{2}{8}$ so I subtracted that from $2\frac{1}{8}$ to get $1\frac{7}{8}$. I composed 2 and $1\frac{7}{8}$ to get $3\frac{7}{8}$. So, $1\frac{3}{4}$ and $2\frac{1}{8}$ is $3\frac{7}{8}$.
Compensation		$2 + 2\frac{1}{8} = 4\frac{1}{8}$ $4\frac{1}{8} - \frac{2}{8} = 3\frac{7}{8}$ $1\frac{3}{4} - 2\frac{1}{8} = 3\frac{7}{8}$	I composed 2 and $2\frac{1}{8}$ to get $4\frac{1}{8}$. I compensated $\frac{1}{4}$, which is $\frac{2}{8}$ so I subtracted that from $4\frac{1}{8}$ to get $3\frac{7}{8}$. So, $1\frac{3}{4}$ and $2\frac{1}{8}$ is $3\frac{7}{8}$.

Addition and Subtraction Fluency 43

Summary

Students should practice each strategy to build greater flexibility in combining multi-digit decimals as well as fractional amounts. Having been encouraged, initially, to use an iconic model (bar model or number line) and having written out and stated each of their steps, children cognitively imprint the process, which will allow them to remember the strategy for a longer period of time and transfer it to similar, but new problems. Have students complete one of the worksheets below twice a month to continue building their number sense.

TEMPLATE: Addition and subtraction strategy practice (multi-digit)

Expression	Doubles	Place value	Benchmark	Compensation
19 + 18				
23 + 34				
26 + 37				
42 + 53				

TEMPLATE: Addition and subtraction strategy practice (decimals)

Expression	Doubles	Place value	Benchmark	Compensation
1.4 + 1.8				
3.2 + 2.4				
1.3 + 2.5				
3.9 + 1.7				

TEMPLATE: Addition and subtraction strategy practice (fractions)

Expression	Doubles	Place value	Benchmark	Compensation
$2\frac{3}{4} + 1\frac{1}{2}$				
$2\frac{3}{8} + 2\frac{5}{8}$				
$1\frac{1}{2} + 3\frac{3}{4}$				
$3\frac{7}{8} + 2\frac{1}{4}$				

TEMPLATE: Addition and subtraction strategy practice

Expression	Doubles	Place value	Benchmark	Compensation

Multiplication and Division Fluency and Flexibility (0–20)

There is a difference between knowing facts versus developing and remembering facts. Our goal is to ensure students are fluent and flexible with numbers. One important idea to keep in mind is that spatial reasoning, the ability to compose, decompose, and rotate shapes, and measurement concepts are the best predictors of future success in math. By incorporating these ideas into learning multiplication and division, we can increase students' abilities to learn and remember their facts and at the same time build early algebraic knowledge.

Fluency is most commonly determined by a measure of a timed fact test at an approximate rate of 3 seconds or less per fact. In contrast, we define flexibility as the ability to solve problems in a variety of ways, use information already known to find unknown facts, and the capability to determine the most efficient method to use when confronted with a challenging problem.

As mentioned earlier, drill and rehearsal techniques can increase students' short-term gains in multiplication fluency. These drill approaches often take the form of flash cards, chants and songs, repetitious games (often computerized) and the frequent use of timed tests as practice tools. However, when fluency is improved by use of drill, there are two concerning phenomena. First, the gains in fluency do not always translate to flexible use of facts in problem solving situations; and second, these gains in fluency appear not to hold over time.

INITIAL IDEAS TO BUILDING MULTIPLICATIVE THINKING

There are a few essential ideas to build students' basic fact knowledge in relation to multiplication and division. First, it is important to use enactive models (physical tools such as cubes), then iconic models (the bar model and number line are the best), and then symbolic models (including verbal language, equations, and algorithms). With this approach to representing multiplicative situations, it is also important for students to look for patterns and relationships and to move away from counting on from each amount. We will provide a progression of how best to introduce and teach multiplication and division facts from 0 to 20.

Skip counting

Somewhere in first or second grade it is important to have students practice skip counting. Typically, we observe students in school practice orally counting up from 0. For example, you will hear them counting together, 0, 5, 10, 15, 20, 25, and so on to some designated number like 50 or 100. This approach is not effective for most students. This strategy is only effective for a few students who have already built a visual representation of number mentally and who notice patterns and/or are adept at remembering sequences (e.g., have well-developed executive functioning abilities). So, it is important to help each student visualize the relationships, observe and discuss patterns, and improve their memories.

Start by asking students to use cubes that connect, and large pieces of paper (sentence strips work great). Typically, the order of introducing skip counting is 2s, 5s, 10s, then 4s, 8s, followed by 3s, 6s, and 9s, and finally 7s. Let's start by skip counting with 2s to explain this process. Ask each student to draw a straight line on their paper (which should be about 2 feet long). Yes, it is important for them to practice drawing a straight line by themselves. If they need a scaffold (to be used just a few times, not indefinitely), let them use a tool like a meter stick. Ask them to place a tick mark on the far left and write 0. Next, have them build a unit of 2 with their cubes. Have them take the 'unit of 2' and set it on the number line so the left side matches up with 0. They should make a tick mark on the right side of the cubes. Have them trace the cubes to make a rectangle or have them lift the cubes and then have them draw the rectangle. (Both skills are necessary for students to build current and future math skills.) Next, have them write the number 2 at the tick mark. They can step back and then say "0, 2", pointing to the 0 and 2. Or some students might find it helpful to say 0 out loud, then "mouth" 1, then say 2 out loud. They should continue this process until they get to 20.

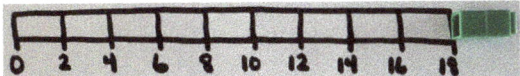

After the bar model (with a number line below) is built, the students should practice counting by 2s from 0 to 20 and back down to 0. Have the student practice the skip counting without looking at the bar model and only look when they pause and get stuck: "0, 2, 4, 6, 8, 10, 12, 14, 16, 18, 20, 18, 16, 14, 12, 10, 8, 6, 4, 2, 0!"

Now it is important to ask students to look for and describe patterns. "What patterns do you notice?" Encourage the student to look for visual and numerical patterns. One of the powerful aspects of the bar model is that you do not have to imagine the relationships but can actually see them.

Ask the student to use their fingers to represent the distance of two units of 2. "How much is this?" "4." "Now, shift your fingers down the bar model without changing their distance. Stop at 6. Where is your second finger?" "10." "What is the distance between 6 and 10?" "4." "Interesting, will two of these spaces always be a distance of 4? Why?" Have the student move to other places like 12 and 16. Yes, it is always true because the distance of two units of 2 is always a distance of 4.

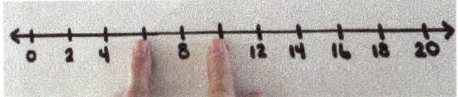

Focus on the ones digits. "When do the digits repeat again?" "The ones digits are 0, 2, 4, 6, 8 and then start over." "Let's try the pattern starting at 20." "20, 22, 24, 26, 28. Yes, it works!" Now switch to look for a pattern in the tens digit. "What happens with the tens digit?" "The tens digit is 0, 0, 0, 0, 0. That is five 0's. Then, there are five 1s for the tens digit, 10, 12, 14, 16, and 18." "What do you think will happen if we continue?" We want students to say, "There would be five 2s for the twenties and five 3s for the thirties."

Put the diagram of 2s to 20 away for now. At two or three different points during the day, ask the student to orally count by 2s from 0 to 20 to 0 without looking at the diagram they built. If they struggle or hesitate, then have them bring the diagram out and use it as a scaffold. Do this for a few days and then introduce a unit of 5. Ask them to build a bar model using a unit of 5 from 0 to 50. You might need to tape two or three pieces of paper together. Ask the same questions as we did with the unit of 2. Then move on to the other numbers suggested above: 10s, 4s, 8s, 3s, 6s, 9s, and 7s.

This process of building a bar model for the numbers 2 through 10, counting forward and backward, and looking for visual and numerical patterns, initiates a strong foundation for seeing and hearing multiplicative relationships. Do not rush this process. This could, and should, take a few months, and students might need to go back and start the entire process over. It might be helpful to rebuild some of the bar models instead of referring to their initial one.

Area models

The next stage in the progression is to enactively build and then draw area models for each of the numbers. Now we are working with two dimensions (where the bar model was linear and essentially one dimensional, despite it having a two-dimensional height we don't measure), so it is cognitively more demanding for students to notice the patterns. Comparing area is simply more difficult than comparing lengths. Look at two tables or desks that are different shapes but similar in size. It is much more challenging, conceptually, to discern whether the top of one table has a greater area than that of another table. However, comparing side lengths of rectangular tables is quite easy to do visually. This is part of humans' innate perceptual measurement of quantity and has historically been the main way we measure attributes of the world around us. Regarding the building of area models for fact fluency, it is important here for students to initially iterate each unit to compose the areas and then to examine relationships by decomposing them.

Let's start examining area models with a unit of 5. Using cubes to build a unit of 5 and then trace around it, leaving room above, to build a bar model. Now that we have a unit of five, we are going to iterate (copy with no gaps or overlaps) the unit above the previous unit three more times for a total of four units of 5. Another way to describe this process is to sweep the unit of 5 up four times. It is a measurement idea of accumulating space, so we are going to label it in 5s from 0 to 20 on the side each time we iterate another unit of 5. Next, have students write an addition equation that matches what they just did. It should look like the following diagram.

 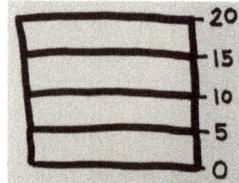

5 + 5 + 5 + 5 = 20

Notice that we are not using graph paper at this time. We want students to see the unit of 5 being iterated and see the 5s. Graph paper can be used as a scaffold later when we are working on different strategies. Now ask the student to look for and describe any patterns they notice. They might say the ones digit is always 0 and 5, which they might have noticed from previous patterns they found when creating the linear bar model. Ask them the following questions. "How many units of 5 are in the model?" "I see four units of 5." "If we compose two units of 5 together as pairs, how much do we have and how many of them do you see?" "Two units of 5 compose 10. And there are three of them." "If we compose three units of 5 together as triplets, how much do we have and how many of them do you see?" "Three units of 5 compose 15 and I see two of them." And if we compose four units of 5 together, how much is that and how many do you see?" "Four units of 5 compose 20 and I only see one of them."

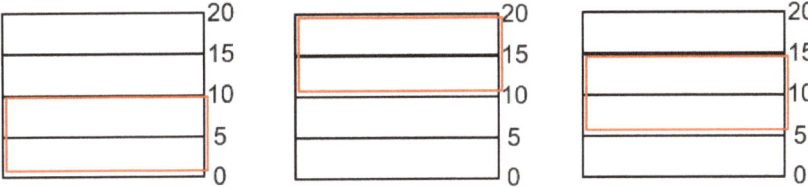

Now we want students to practice writing multiplication notation. Traditionally, you might have learned that the **multiplier** times the **multiplicand** gives you the product. The etymology of these words actually helps us learn multiplication of whole numbers, and of rational numbers like fractions and decimals, in a more conceptual manner. The multiplier is the number of times you are going to iterate the multiplicand, which is the quantity or unit we are working with. So, using these structural words, we can say multiplication is the iteration of a unit to get the total. In our case, we have four iterations of a unit of 5 to get 20. We write this as 4 × 5 = 20. This order matches the common method in the United States of describing 4 × 5 as "4 groups of 5." Using the terms iteration and unit are simply more formal and precise.

Multiplication is **commutative** which means 4 × 5 and 5 × 4 both have a product of 20. Because of this, we tend to use the word "factor" for both the multiplier and multiplicand. Abstractly, it doesn't matter how we write the factors (4 × 5 or 5 × 4). However, it is important to note when describing realistic and visual situations, that each number distinguishes between the unit and the amount of times it is being iterated. So, even though a multiplication statement can be written either way, we will be specific, when describing a realistic or visual situation, about which factor we are using as the amount or quantity being multiplied (the multiplicand), and which factor we are using as the number of iterations (the multiplier). These relationships become increasingly clear when you move to multiplying whole numbers with fractions. It is quite easy to think of 4 × $\frac{1}{3}$ as being a unit of $\frac{1}{3}$ iterated four times. However, modeling and describing $\frac{1}{3}$ × 4 is quite a bit more complex. In this second number set, you are actually finding $\frac{1}{3}$ of a unit of 4. While the products are the same, there are enough differences that many schools in the U.S. reserve 4 × $\frac{1}{3}$ for Grade 4 students and wait until Grade 5 to teach $\frac{1}{3}$ × 4. If you're curious why the distinction would matter, a short answer is that students learning to multiply and divide fractions are progressing into concepts of ratio and proportion. Reasoning proportionally about why $\frac{1}{3}$ × 4 actually *decreases* the size of the unit of 4 down to $\frac{4}{3}$ will be invaluable in secondary mathematics when students learn concepts of similarity, geometry, scale, and transformations.

Just as we had the student build a bar model for the different numbers from 2 to 10, we are going to have them now build an area model. Have them look for patterns, but this time encourage them to look for relationships by decomposing each area model. They should build area models for 2s, 5s, 10s, then 4s and 8s, then 3s, 6s, and 9s and finally 7s. Have them iterate each unit 10 times.

Here is an example for 4s. First, build a bar model for a unit of 4. Then, iterate that unit of 4 ten times, labeling the left side 0 to 10 to represent the iterations and the other side from 0 to 40 to represent the total. It should look similar to the figure below. Here is a possible conversation. "What does each rectangle represent?" "4." "Show me 8." "It's two of the rectangles." "Do you see an 8 somewhere else in the area model?" "Yes, I see one here in the middle." "Where does it start and end?" "It starts at 24 and ends at 32." What is the difference between 32 and 24?" "It's 8." "Show me some other examples." "It's everywhere, 8 to 16, 32 to 40 and 20 to 28." "What do you notice about the ones and tens digits? What patterns do you notice?" "The ones digit goes 0, 4, 8, 2, 6 and then starts over. And the tens digit goes 0, 0, 0, 1, 1, 2, 2, 2, and then 3, 3. I think it would continue to have 3 of the same tens digit and then 2 of the same. So, the next tens would be 4, 4, 4 and then 5, 5."

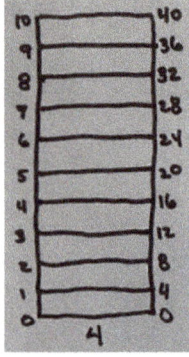

Bar Model of 4's

After the student has explored some patterns, it is time to work on composing different products using the area model. Let's examine some different relationships. One question to ask is which numbers are good benchmark numbers to make composing different amounts of 4 much easier? Typically, it would be counting by two 4s, four 4s (a square number), and five 4s. So, now, have the student build a few of each of these products by having them cut out a piece of paper that fits over 1 × 4, 2 × 4, 4 × 4, and 5 × 4 and then label each piece of paper with the equations. It should look like the figure below.

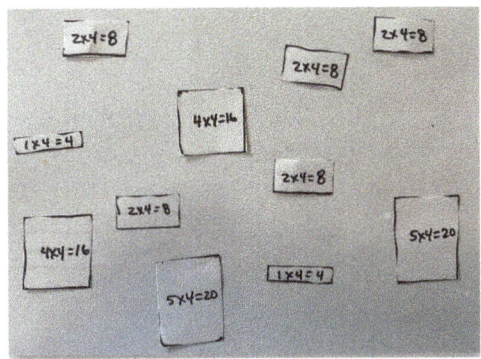

Now, we are going to use these products to compose larger products of 4s using the area model and equations. Have the student build the following area models and determine the products:

2 × 4 + 2 × 4

2 × 4 + 1 × 4

4 × 4 + 4 × 4w

5 × 4 + 1 × 4

5 × 4 + 4 × 4

5 × 4 + 2 × 4.

Their area models should look similar to the ones below. Some students may need to use graph paper as a scaffold to initially engage with this task.

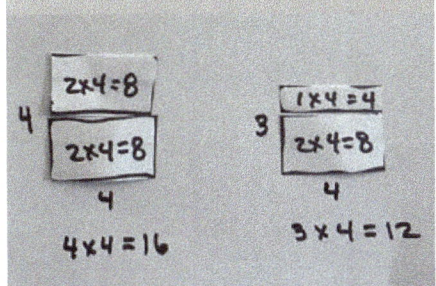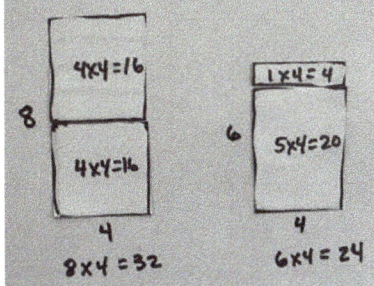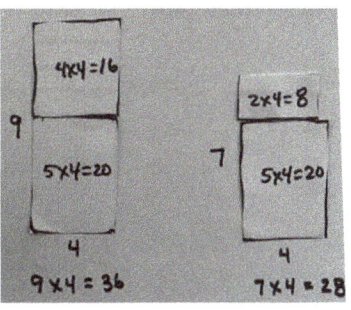

After completing these products, it is important for the student to explain the process. Here is the sentence frame they can use. "I know 2 times 4 is 8, and 2 times 4 is 8. And I know 8 plus 8 is 16. So, 4 times 4 is 16." Having students say the partial products[14] and then the final product out loud reinforces these different facts and starts to build flexibility and a conceptual model of multiplication. Each day or every other day, have students work on this process with different units.

Now that students have composed different products, it is important for them to decompose an array into different products. This is not conceptually easy for many students and is a task that needs to be practiced. Building on a similar theme as earlier, provide them with different arrays visually, in this order of priority: 2s, 5s, 10s, 4s, 8s, 3s, 6s, 9s, and 7s. Ask them to decompose them into two arrays and write the partial products.

Here is an example with 6s. Use the template that includes the open area models and have the student create 3 of the same open area model by tracing over the template. Then ask, "I want to know what 8 × 6 is. Decompose the 8 into 5 and 3, 4 and 4, and 6 and 2. How would you construct each of these partial products?" See a student's work below. Ensure they show how the 8 was decomposed, similar to how this student represented it on the left side. Then ask the student to state the entire process using the following sentence frame: "I know __ times __ is __. I also know __ times __ is __. And __ plus __ is __. So, __ times __ is __." For 5 × 6 and 3 × 6 it would be: "I know 5 times 6 is 30. I also know 3 times 6 is 18. And 30 and 18 is 48. So, 8 times 6 is 48." Now have students go through all the different numbers over the next few weeks.

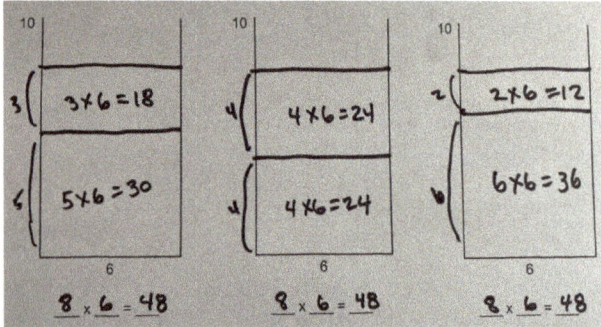

---

14 Partial products refer to all the products used to determine the original product. So, to find 7x8, I might use 7x5=35 and 7x3 =21. These are named partial products to find the actual product: 7x8=56.

TEMPLATE: Decomposing numbers 2 to 10 using an open area model

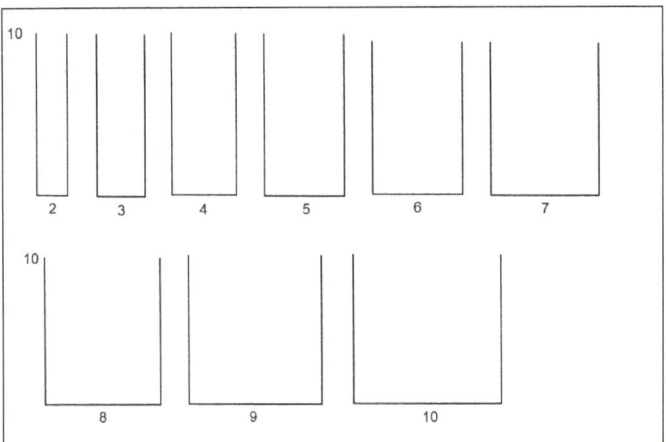

MULTIPLICATION AND DIVISION STRATEGIES

There are four strategies to build multiplicative[15] fluency and flexibility: **doubles**, **squares**, **benchmark**, and **compensation**. The goal is to build a strong foundation with multiplication, so that students remember facts but, more importantly, are able to use them in mathematical and real-world situations, from proportional reasoning to functions, to data analysis. We will again highlight Bruner's cognitive processes of coming to understand topics by going through iconic and symbolic representations. Before a student works on these strategies, make sure they have gone through the earlier models and language exercises, skip counting, discussing patterns, composing and decomposing partial products with area models, so that they are ready to learn, practice and use flexible multiplication strategies.

Doubles strategy

The Egyptians were known for using the process of doubling to find different multiplication and division facts. There is something cognitively appealing to students (and adults) about doubling. Here we introduce the ratio table[16] as a way of keeping track of multiplicative situations. First, create an area model that proportionately represents the area of the quantity we are doubling. Second,

---

15 'Multiplicative' describes any situation that involves multiplication. This includes multiplication by a fractional amount as well, and therefore includes division. For example, multiplying something by $\frac{1}{4}$ is equivalent to dividing it by 4.

16 We first learned of the wide use of the ratio table to keep track of and solve multiplicative situations in the Netherlands in the 1980s. We used the ratio table in designing a new 5-8 curriculum series in the 1990s called Mathematics in Context.

create a ratio table to keep track of these products. And finally, use combinations of these different products to create larger new products.

Doubles strategy for multiplication

Take the example of doubling the quantity 3. We construct an area model that represents sixteen threes. One way to keep track of the products is to construct a unit of 3 with a vertical line and then iterate or sweep it out horizontally to the right 2, 4, 8 and 16 times, keeping track by labeling these amounts on the top horizontal line. Then have students keep track of the total on the bottom horizontal line. The area model should be as proportionate as possible. It is best to attempt drawing the area model by hand on white paper. If a scaffold is needed, then have students use graph paper for a few weeks and then attempt it with blank paper again.

Figure. Doubles models for 3 (bar model and ratio table)

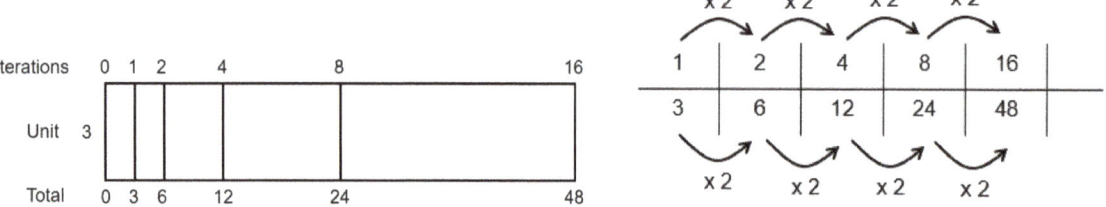

While the area model is an iconic or visual model, we can also keep track of the products using a ratio table, which is a symbolic model. To do this, have the students construct a horizontal table and then keep track of the products. It is beneficial to have the student represent which ratio or composed unit they used (in this example a unit of 3) and what they multiplied by. In this case we continued to multiply each previous unit by two. These models are increasing horizontally as opposed to some previous examples that grew vertically. The examples below are more common in the elementary grades. The previous vertical examples are helpful for early multiplication learners because students are intuitively interested more in the *height* than in the *width* of objects. Think of young children comparing heights to see who is tallest. Additionally, the vertical models will re-appear in later grades when students work with the coordinate plane. Flexibility in model construction and orientation is important to develop, so both presentations should be used with students over a period of years.

Once we have constructed these mathematical models (area model and ratio table), we can use them to find all the other products for 3. For example, if we want to know 6 × 3 or 13 × 3, we can combine the composed units or doubles. For 6 threes, I can compose 4 threes and 2 threes, which is 12 and 6, or 18. So, 6 times 3 is 18. If I want to know how much is 13 threes, I can compose 8 threes, 4 threes, and 1 three, which is 24, 12, and 3, making 39. So, 13 times 3 is 39.

Figure. Area model and ratio table example for 6 × 3.

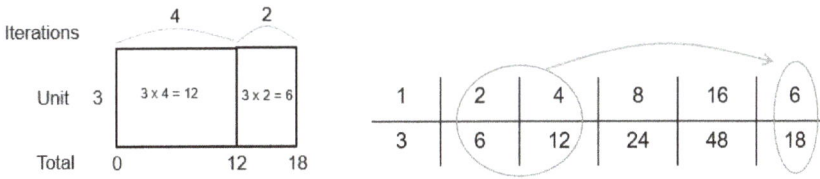

Figure. Area model and ratio table example for 13 × 3.

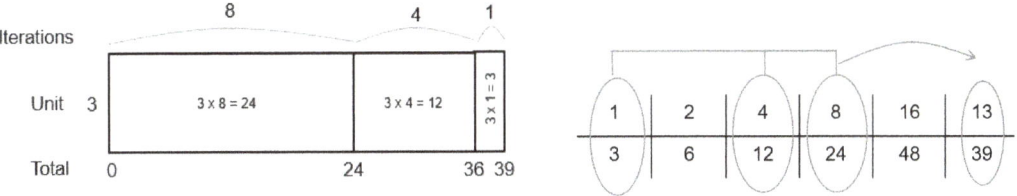

Once the reasoning is understood, students will need to practice. Have them create an area model, use a ratio table, write the partial products with equations, and finally explain in writing (and orally) how they used the doubles strategy to find the product. Start with the expressions listed in the template.

TEMPLATE: Doubles strategy

Use the doubles strategy to complete the following. Create an area model, write equations to match, and then use this sentence frame to describe the process. "I know __ times __ is __. I also know __ times __ is __. So, __ times __ is __."

Expression	Area model	Sentences
9 × 8	Iterations 0, 1, 2, 4, 8; Unit 9; Total 0, 9, 18, 36, 72	I know 9 times 2 is 18. And 18 plus 18 is 36. So, 9 times 4 is 36. And 36 plus 36 is 72. So, 9 times 8 is 72.
	Ratio table	**Equations**
	×2, ×2, ×2 across: 1, 2, 4, 8 / 9, 18, 36, 72	9 × 2 = 18 9 × 4 = 36 9 × 8 = 72
Expression	**Area model**	**Sentences**
4 × 6		
	Ratio table	**Equations**
Expression	**Area model**	**Sentences**
3 × 8		
	Ratio table	**Equations**
Expression	**Area model**	**Sentences**
6 × 7		
	Ratio table	**Equations**
Expression	**Area model**	**Sentences**
7 × 9		
	Ratio table	**Equations**

Doubles strategy for division

The doubles strategy can also be used to build and solve division facts. Because multiplication and division are inverse operations, we can think about partitioning instead of iterating. However, our brain is really using multiplication facts while dividing. For example, we can think of 6 × 7 as either 6 iterations of the unit 7 (7, 14, 21, 28, 35, **42**) or 7 iterations of the unit 6 (6, 12, 18, 24, 30, 36, **42**). If someone asks us what 42 ÷ 7 is, they are asking how many 7s fit into 42 or how many times we would partition 42 to get 7. Instead of counting back by 7s, we can use the doubles strategy.

Take 63 ÷ 7 as a second example. First, let's construct an open area model with 7 as our unit. Unlike with multiplication, we don't know the other factor yet, but we know the total 7s will be 63. Next, we build the area model and the ratio table by doubling until we get close to 63. One 7 is 7. Two 7s are 14. Four 7s are 28. Eight 7s are 56. If we double again, we shall be past 63. At this point we look for a combination of the double 7s to compose 63 using 56. Notice that one more 7 would be 63 (56 + 7), so that would be nine 7s. Always end by restating the fact. "I know 63 divided by 7 is 9."

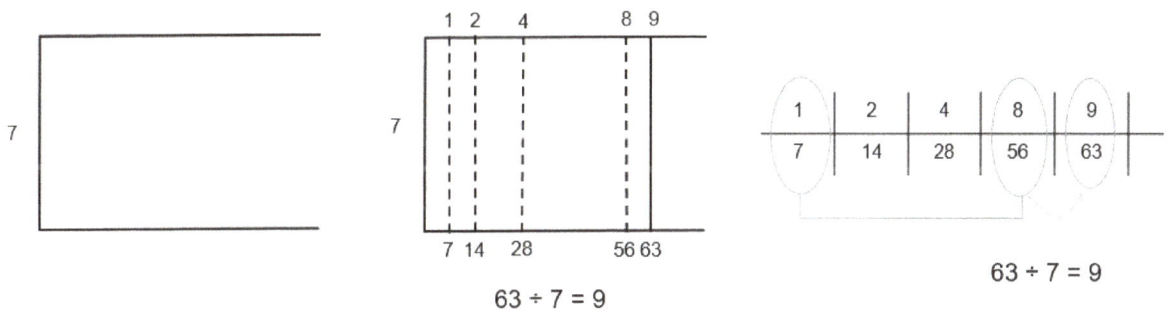

Summary

The doubles strategy is a good first approach to building multiplication and division fluency. Similar to the ancient Egyptians, we like doubling patterns. It is important to have students initially create area models to build a visual representation that is proportional[17] and then keep track of the partial products with the ratio table, a more abstract but flexible model. It is equally important for students to state the process out loud. Both of these approaches help build a strong multiplicative schema for the student to draw on later.

---

17 We continue to stress the importance of developing a visual basis and understanding of proportion along with equations. Proportional reasoning is paramount for algebra and higher levels of mathematics and with real-life situations.

Squares strategy

A second multiplicative strategy uses squares. As with doubles, humans are attracted to square numbers. When you multiply a number by itself, for example, 4 × 4, it is written as 4^2, which is 4 to the second power. The exponent (e.g., 2) tells us how many times to multiply the base (e.g., 4) by itself. However, because the visual model for 4^2 is a square, we typically refer to this quantity as 4 squared. So, we will go with this more informal way of describing this relationship.

Write down all the squares from 0 to 10. Ask the student to say them out lout and then look for patterns.

$0^2 = 0$ $1^2 = 1$ $2^2 = 4$

$3^2 = 9$ $4^2 = 16$ $5^2 = 25$

$6^2 = 36$ $7^2 = 49$ $8^2 = 64$

$9^2 = 81$ $10^2 = 100$

Students might observe that the differences between the consecutive pairs of square numbers are 1, 3, 5, 7, 9, 11, 13, 15, 17, and 19. Why does that pattern occur? Draw an area model of 1^2 using graph paper and then use that area model to build 2^2 and so on. Look at the difference.

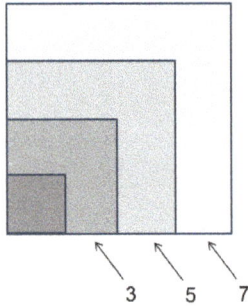

Squares strategy for multiplication

When using a square strategy for multiplication, we are decomposing one of the numbers. We find the square of the smaller number and then we find the remaining amount by decomposing the other factor. Let's try two examples. First ask, how we might use a square number for 5 × 4. Then ask the same questions for 8 × 6. For each, draw an area model, write equations to match, and then use this sentence frame to describe the process. "I decomposed the __ into a __ and a __. Then, I multiplied __ times __ to get the square number __. Then, I multiplied __ times __ to get __. I composed __ and __ to get __. So, __ times __ is __."

For 5 × 4:

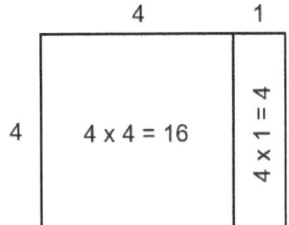

4 × 4 = 16

1 × 4 = 4

16 + 4 = 20

5 × 4 = 20

I decomposed the 5 into a 4 and a 1. Then, I multiplied 4 times 4 to get the square number 16. Then, I multiplied 1 times 4 to get 4. I composed 16 and 4 to get 20. So, 5 times 4 is 20.

For 8 × 6:

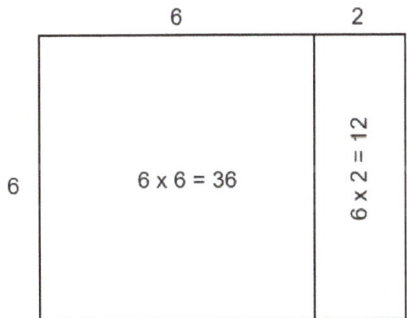

6 × 6 = 36

2 × 6 = 12

36 + 12 = 48

8 × 6 = 48

I decomposed the 8 into a 6 and a 2. Then, I multiplied 6 times 6 to get the square number 36. Then I multiplied 2 times the 6 to get 12. I composed 36 and 12 to get 48. So, 8 × 6 is 48.

The difficulty tends to be in knowing what to multiply when the one factor is decomposed. Many students will state 6 times 6 is 36, but then they are not sure whether they should multiply 2 times 8 or 2 times 6. This is why the visual (area model) is so important, as is the language of describing 6 iterations of 6 and 2 more iterations of 6 to get a total of 8 iterations of 6. Mathematically, we are using the **distributive** property: 8 × 6 = (6 + 2) × 6 = 6 × 6 + 6 × 2 = 36 + 12 = 48. Older students can write out and discuss the amounts using this notation.

Squares strategy for division

For division, we will work in the opposite direction and look for the closest square number. As a warm-up, have students practice counting the first ten or twelve square numbers (forward and backward): 1, 4, 9, 16, 25, 36, 49, 64, 81, 100, 121, 144.

Let's work on an example: 35 ÷ 5. Start by creating an open area model, labeling one side with the divisor (the unit 5) and find the square for the nearest square that is just smaller than the dividend (or total amount). In this case the dividend is 35. So, without going over 35, the next closest square would be 25, because 36 (the next square) would be too large. So, we know 25 ÷ 5 = 5. We have iterated five 5s to get 25. That leaves a quantity of 10 to compose the dividend (35 – 25). How many more units of 5 fit into 10? 2 more. That means seven 5s fit into 35. So, 35 ÷ 5 = 7.

For 35 ÷ 5

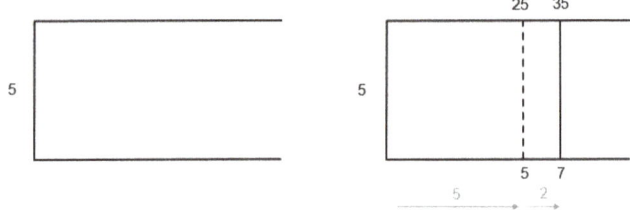

I know 25 ÷ 5 = 5.

And I know 10 ÷ 5 = 2.

So, 35 ÷ 5 = 7

Mathematically, this works using the distributive process.

    35 ÷ 5 =

    (25 + 10) ÷ 5 =

    (25 ÷ 5) + (10 ÷ 5) =

    5 + 2 = 7.

Multiplication and Division Fluency and Flexibility (0–20)

TEMPLATE: Squares strategy

Use the squares strategy to complete the following. Create an area model, write equations to match, and then use this sentence frame to describe the process. "I know __ times __ is __. I also know __ times __ is __. So, __ times __ is __."

Expression	Squares Area Model and Equations	Sentences
4 × 5	(Area model: 4 × 4 square with 4 × 1 = 4 strip; total width 4 + 1, height 4) 4 × 4 = 16 4 × 1 = 4 16 + 4 = 20 4 × 5 = 20	I know 4 times 4 is 16. I know 4 times 1 is 4. So, 4 times 5 is 20.
5 × 8		
6 × 7		
9 × 8		
12 × 8		

Benchmark strategy

The benchmark strategy will typically be the most robust and used strategy for finding a product. It is sometimes described as using derived facts, which means using two known facts to determine the product. When students practice this approach their ability to remember and use other facts increases. The more practice students have deriving facts or using the benchmark strategy, the more fluent they become.

The benchmark strategy is the process of decomposing one of the factors into two friendlier numbers where one is a benchmark number. A benchmark number is typically a multiple of 5 or 10. Then, these two decomposed numbers are used to find two known facts. These two facts are added together to determine the initial product. The most common benchmark facts will be using multiples of 5 or 10.

Benchmark strategy for multiplication

Let's try an example: 6 × 7. We can decompose either the 6 into a 5 and a 1 or the 7 into a 5 and a 2. We will show what both look like in an area model and how we should write the equations and describe the process.

I can decompose the 7 into a 5 and a 2. Then I iterate the six 5 times, sweeping to the right, to get 30. Then, I still have two more 6s, so I iterate the six 2 more times to the right to get 12. I compose the 30 and the 12[18] to get 42. The final product of 6 times 7 is 42.

6 × 5 = 30

6 × 2 = 12

6 × 7 = 42

---

18 The student should continue to practice their addition strategies to add. In this case, 30 + 10 + 2 or 30, 40, 42.

Alternatively, I decompose the 6 into a 5 and a 1. Then, 7 is iterated 5 times, sweeping up, to get 35. I have 1 seven left over, or 1 iteration of 7 on top. I compose or add the 35 and the 7 to get 42. If I had iterated the 7 six times, I would also have got 42.

	7
1	1 x 7 = 7
5	5 x 7 = 35

$5 \times 7 = 35$

$1 \times 7 = 7$

$6 \times 7 = 42$

In these descriptions, notice that we intentionally did three things. First, we went back to describing the iteration of a unit. After working on the doubling and square strategies, some students get confused as to what they are decomposing and then what number is being iterated. So, having the student describe the iteration or the sweeping of a unit helps reinforce what multiplication is. Second, when composing the two derived facts, we describe how we would add the products in our head by decomposing one of the addends. This reinforces flexibility with addition — a bonus. Third, when we line up the equations for the partial products we keep the number being decomposed in the same order. So, the 5, 2 and 7 are lined up in the first example as are the 5, 1 and 6 in the second example. This helps keep track of the process as well and can prevent students less comfortable with this process from becoming lost in the various decompositions of the factors.

As in the earlier example, this strategy can be described mathematically as decomposing a number and using the distributive property.

6 × 7 =

6 × (5 + 2) =

6 × 5 + 6 × 2 =

30 + 12 = 42

6 × 7 = 42

Benchmark strategy for division

We can also use the benchmark strategy if we are working on division facts. If we are solving 42 ÷ 6, then we would create the open area model with the unit 6 to be iterated out to 42. We use our benchmark of 5 to determine whether we can do that many iterations. Yes, 5 iterations of 6 get us to 30, which is still less than 42. The difference between 42 and 30 is 12. Ask how many 6s fit into 12. Two. We iterate the unit of 6 two more times to get to 42. We found 7 total iterations. So, 42 ÷ 6 is 7.

 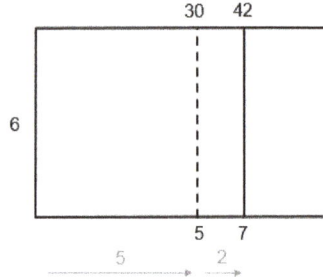

Multiplication and Division Fluency and Flexibility (0–20)   65

TEMPLATE: Benchmark strategy

Use the benchmark strategy to complete the following. Create an area model, write equations to match, and then use this sentence frame to describe the process.

"I know __ times __ is __. I also know __ times __ is ___. So, __ times __ is ___."

Expression	Benchmark area model and equations	Sentences
9 × 7	Total 0, 45, 63; 9 × 5 = 45; 9 × 2 = 18; Iterations 5, 2. $9 \times 5 = 45$ / $9 \times 2 = 18$ / $45 + 18 = 63$ / $9 \times 7 = 63$	I know 9 times 5 is 45. I also know 9 times 2 is 18. So, 9 times 7 is 63.
6 × 8		
6 × 9		
8 × 8		
8 × 7		

Compensation

The fourth strategy to introduce to students is compensation. This strategy looks similar to the benchmark strategy in the sense that you round one of the factors to a benchmark number. But, whereas in the benchmark strategy you are decomposing one of the factors into a benchmark number, here you are adding a number to change a factor and, therefore, must compensate for that amount later.

Compensation strategy for multiplication

Here is an example of the compensation strategy for 8 × 9. Notice how invaluable visual models are when working with compensation. The additional iteration of 8 to change 8 × 9 to 8 × 10 is seen in the model. Using only symbolic notation can be difficult with compensation at the early stages of learning the strategy.

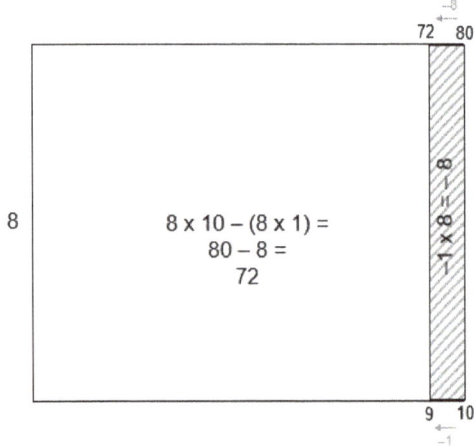

8 × 9

8 × 10 = 80

− (8 × 1) = − 8

80 − 8 = 72

8 × 9 = 72

I know 9 is close to 10 and I know 10 × 8 is 80. I did 1 iteration of 8 too many, so I need to compensate by subtracting 1 × 8 or 8 from 80 to get 72 (or adding -1 x 8, or -8). 9 times 8 is 72.

Students do have trouble initially keeping track of whether they subtract 8 or 9. If students try this and get 71, remember to have the student refer back to the area model, an iconic model, and line up the equations to help keep track of, in this case, the 9, 10, and –1. Then, have students describe the process. For example, "I need nine 8s, but I am iterating ten 8s to get 80. As I iterated one 8 too many, I need to subtract one 8 to get 72. So, I know 8 times 9 is 72."

This strategy is used in algebra and higher levels of mathematics as a really important way of solving more complex problems. It makes a difficult calculation easier by modifying the expression to create a set of easier relationships. Mathematically, we use the additive identity property to add 0 to a number, knowing it won't change the number: 9 + 0 = 9. Then, we use the additive inverse property to add a number with its inverse (opposite): 1 + -1 or 1 - 1 = 0. The final step uses the distributive property. So, algebraically the process looks like this:

$9 \times 8 =$

$(9 + 0) \times 8 =$ Additive identity property

$(9 + 1 - 1) \times 8 =$ Additive inverse property

$(10 \times 8) - (1 \times 8) =$ Distributive property

$80 - 8 = 72$ Compensation

$9 \times 8 = 72$ Initial product

Have students use compensation with only one factor at this point. When we discuss working with larger numbers, we will show how compensation works with, for example, double-digit numbers, and we will consider its connections to algebra.

Compensation strategy for division

Compensating with division is more difficult, but it can be used. Multiply the divisor by a benchmark number (e.g., 5 or 10) that creates a product just greater than the dividend. Then subtract this compensated product from the dividend. This may be a strategy that is explored more than mastered.

Here is an example with 72 ÷ 8. I multiply 8 times 10 to get 80. I know 80 ÷ 8 = 10. I only had 72, so 80 take away 72 is 8. And 8 divided by 8 is 1. I used 1 unit of 8 too many so I need to compensate. 10 minus 1 is 9. So, 72 divided by 8 is 9.

MATH FACTS

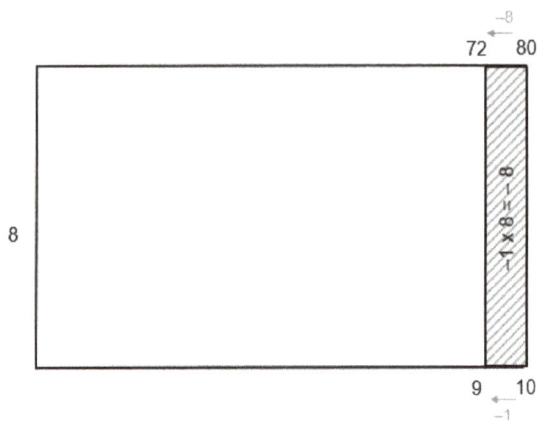

72 ÷ 8 = ?

80 ÷ 8 = 10

80 − 72 = 8

−8 ÷ 8 = −1

72 ÷ 8 = 9

TEMPLATE: Compensation strategy

Use the compensation strategy to complete the following. Create an area model, write equations to match, and then use this sentence frame to describe the process. "I know __ times __ is __. I also know __ times __ is __. So, __ times __ is __."

Expression	Compensation area model and equations	Sentences
9 × 6		
8 × 8		
4 × 9		
7 × 8		

Practice and probes

Now that all four strategies for multiplication and division have been introduced, it is important to practice these. Each of the strategies provides a different way of thinking about multiplicative situations and builds fluency and a solid foundation to move on to multi-digit multiplication. The point of this is to build flexibility, and not to master all four strategies for all multiplication facts. From working on these strategies, students will eventually use them when needed.

Practice 1: All four strategies

Add four expressions to the template below and then have the student (over two weeks) complete each strategy, although doubles typically would only be used when the numbers are close together. The template can be completed for the area model only, equations only, or both. Once completed, the student should randomly toss a chip or roll a die on to the chart to land on different cells and explain the strategy out loud.

TEMPLATE: Multiplication strategy practice

Expression	Doubles	Squares	Benchmark	Compensation

Practice 2: Strategy cards

As with traditional flash cards, a student is presented with a multiplication expression on one side, while on the back are two or three strategies that they might want to use to solve the expression. The expression is written horizontally[19] to encourage multiple strategies and to be similar to algebraic expressions.

---

[19] When a student sees the expression set out vertically, they tend to think of using the traditional algorithm. Initially, we want to ensure they are thinking strategically to build a stronger foundation.

Students are shown the expression and can be asked what the product is. To build a deeper fact base and to build more flexible thinking, encourage the student to explain how they could use one of the strategies to determine the product. Or (more challenging) ask the student to explain one of the strategies on the back of the card. For instance, for 6 × 9, you can ask the student what a compensation strategy would be.

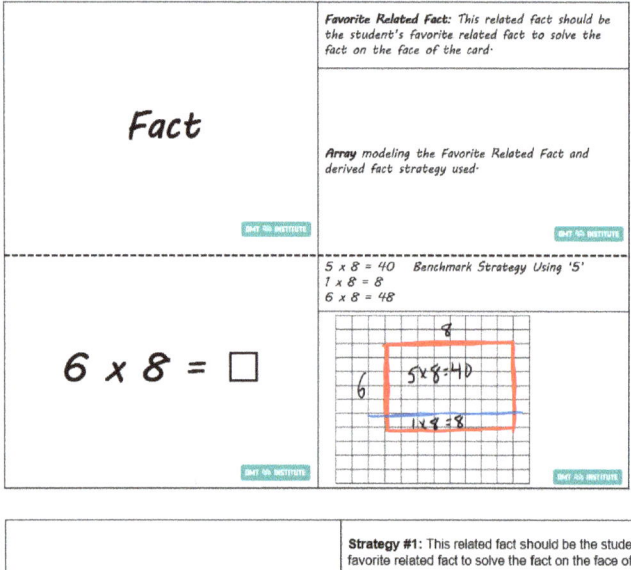

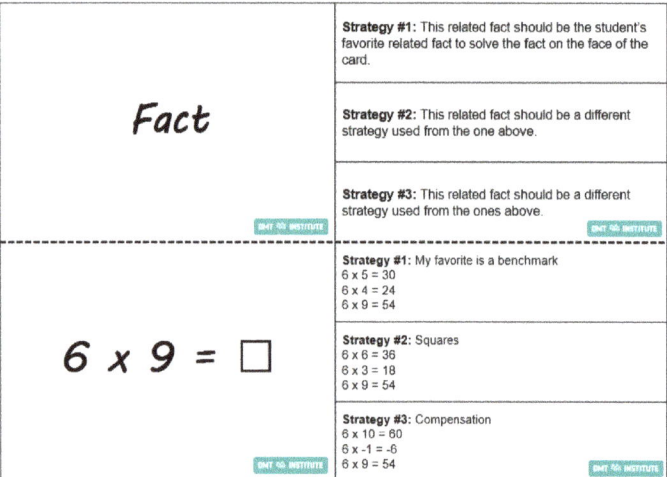

Example

Use the following template to build your own. Or you can go to www.dmtinstitute.com and download a pdf with these facts already on the cards.

Template: Strategy cards

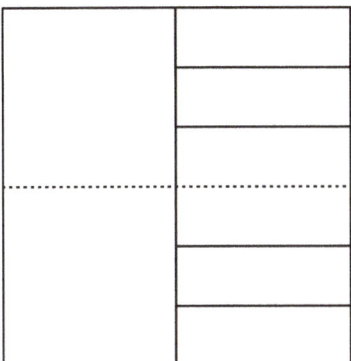

Practice 3: Multiplication matrix and arrays

The matrix of number products shown below can be used as an activity. Give students a blank template and have them fill out the entire matrix, then ask them to look for patterns. Have them put wax paper or patty paper over the matrix or insert it in a plastic sheet. Have them use overhead markers to draw an array on the matrix for a fact with which they are having trouble. They can find smaller arrays and facts within that array. Use the arrays within the more difficult array and write down the smaller facts to create the whole product using clear plastic sleeves (or sheet protectors) — students can draw in smaller arrays to help build the larger or more difficult array with partners.

12	12	24	36	48	60	72	84	96	108	120	132	144
11	11	22	33	44	55	66	77	88	99	110	121	132
10	10	20	30	40	50	60	70	80	90	100	110	120
9	9	18	27	36	45	54	63	72	81	90	99	108
8	8	16	24	32	40	48	56	64	72	80	88	96
7	7	14	21	28	35	42	49	56	63	70	77	84
6	6	12	18	24	30	36	42	48	54	60	66	72
5	5	10	15	20	25	30	35	40	45	50	55	60
4	4	8	12	16	20	24	28	32	36	40	44	48
3	3	6	9	12	15	18	21	24	27	30	33	36
2	2	4	6	8	10	12	14	16	18	20	22	24
1	1	2	3	4	5	6	7	8	9	10	11	12
X	1	2	3	4	5	6	7	8	9	10	11	12

TEMPLATE: Multiplication matrix

Probe 1: Two-minute probes

About once a month, a student can take a multiplication or division probe (www.dmtinstitute.com/fluency-probes). These probes take two minutes each. The calculations are randomly selected from a set bank of items, and the test provides an accurate score of fluency. Only test students when they have had the opportunity to practice the different strategies. You can track the number of items attempted, the percentage correct and the average time taken per item.

Probe 2: Multiplication matrix

You can also use a blank copy of the multiplication matrix above as a probe. Set a timer for two minutes and have the student try to complete the multiplication matrix. Have them check how many of the responses they got correct. Ask the student what patterns they notice in how they were filling out the table. Did they put in all the squares? Did they start with 1s, then 2s, and so on? Did they go horizontally or vertically? Some students fill out the 1s, 2s, 5s, and 10s. Check for growth over time.

Multiplication and Division Fluency and Flexibility (Multi-digit and Rational Numbers)

This chapter explores the challenges educators face when developing and justifying their reasoning strategies for multi-digit multiplication. In order to teach multiplication effectively, adults need to possess a solid understanding of whole number multiplicative computation and be able to reason about numbers and operations, especially since schools hold students to these same expectations. Therefore, teachers and family educators must develop their ability to reason about numbers and operations, in order to better equip themselves to support children to do the same. This goal can be achieved when an emphasis is placed on mathematical justification through the development of alternative strategies for whole number computation, based on reasoning about multiplication (reasoning strategies) and justifying the validity of these strategies. Thus, educators have opportunities to deepen their own understanding in order to support students. This approach also presses learners to apply the meaning of multiplication (or any operation) as the foundation of their reasoning. When educators simply rely on rule-based mathematics and computation, they automatically use these to both solve and explain their answer, leading to numerous challenges in developing reasoning strategies and their associated justifications.

Developing an understanding of multi-digit multiplication

Computation taught without meaning leads to poorly remembered procedures, errors, and difficulty with retention and transfer.[20] Multidigit multiplication is a case in point; the traditional algorithm can disguise the meaning of the numbers and sacrifices scaffolds for problem solving in order to be efficient and space saving.[21] As a result, students may make errors in following the correct sequence of steps, or in recording numbers, or they may struggle to apply the procedure appropriately in problem-solving tasks.

To develop understanding, computational knowledge needs to build upon intuitive knowledge. Researchers have found that students use intuitive knowledge and invented strategies to solve multiplication problems before formal instruction takes place; students in stages as early as kindergarten have demonstrated this ability.

Substantive knowledge can develop directly out of intuitive strategies. Students use invented models that support various strategies to solve multiplication problems when these are presented in realistic contexts. Working with different types of problem-solving situations to draw on intuitive knowledge, educators can direct students to more formal iconic models such as number lines, bar models, or arrays. Additionally, intuitive multiplication strategies can lead students to using tools such as ratio tables or methods such as partial products that will ultimately illuminate some of the underlying mathematical principles that allow standard algorithms to work.

After students have spent some considerable time practicing the different fluency strategies we have previously described with smaller numbers, they can move onto multi-digit whole numbers and can use the strategies with rational numbers such as fractions and decimals. This chapter will examine how the strategies can be used in tandem to solve these types of problems. Working on these problems builds flexibility with not only multiplicative situations, but additive scenarios as well due to the need to compose and decompose numbers throughout each method.

Fluency and flexibility with multi-digit multiplication using integers

We are going to examine four expressions: 18 × 19, 32 × 24, 29 × 42, and 47 × 58. The first expression involves two quantities less than 20 with ones values greater than

---

20 Brownell, W. A. (1947). "The place of meaning in the teaching of arithmetic." *The Elementary School Journal* **47**(5): 256-265.
21 Fuson, K. C. (2003). "Toward computational fluency in multidigit multiplication and division." *Teaching Children Mathematics* **9**(6): 300-305.

five. The next three expressions contain numbers greater than 20. The second expression has ones values less than five. The third expression has one number with the ones value greater than 5, and the fourth expression has both ones values greater than 5. Each of these expressions poses different challenges.

Have the student discuss what strategies that they might use. Then, have them find the product by creating a visual model (area model), using equations, and describing their process in written or oral form. Keeping track of the different partial products in a ratio table is always useful and helps build proportional thinking. We will provide different examples.

18 × 19

For 18 × 19, I could use the doubles strategy because 18 is composed of the 16 (2, 4, 8, 16) and 2, but I'm not sure I want to double 19. I think the benchmark strategy can be used by decomposing the 19 into 10, 5 and 4. But the compensation strategy seems to be easiest by compensating 19 to 20.

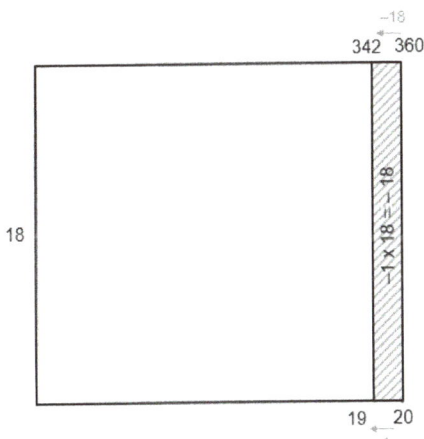

1	10	20	19
18	180	360	342

18 × 19

18 × 20 = 360

− (18 × 1) = −18

360 − 18 = 342

18 × 19 = 342

I compensated 19 to 20. Then, I multiplied 18 times 10 to get 180, then doubled 10 to 20 and 180 to 360. I only wanted nineteen 18s, so I needed to subtract one 18. As 18 is close to 20, I could use compensation by subtracting 20 from 360 to get 340 and then add the 2 back to get 342. So, I know 18 times 19 is 342.

A student might wonder whether compensation can be used for both numbers. Let's examine this strategy. The visual model will help the most. The first step is to compensate both 18 and 19 to 20. So, 20 times 20 is 400. Next, we need to compensate back and subtract out the amounts. From 19 we subtract out 1 (vertical) strip of 20. Then from 18, we subtract out 2 (horizontal) strips of 20 or 40. We subtract the 60 from 400 to get 340. However, you notice that this method subtracts the top right corner twice. So, we have to add back that amount, which is 1 times 2 or 2. We add this to 340 to get 342. So, 18 times 19 is 342.

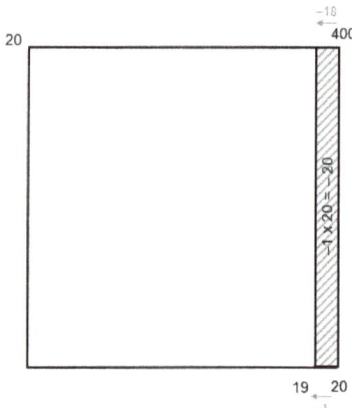

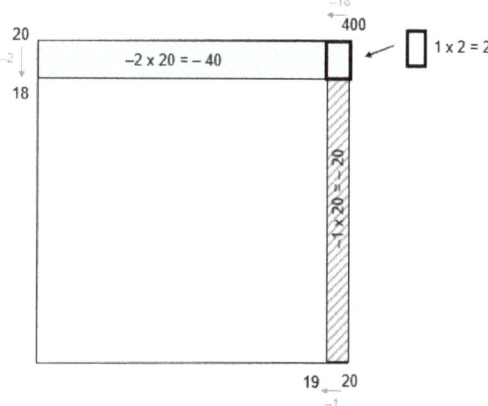

18 × 19

20 × 20 = 400

− (20 × 1) = − 20

− (2 × 20) = − 40

400 − 20 − 40 = 340

1 × 2 = 2

340 + 2 = 342

18 × 19 = 342

We could have also multiplied 20 times 20 to 400 and then subtracted out one 18 and two 19s and then subtracted the small leftover 1 by 2 rectangle. This would be 400 − 18 − 38 − 2 = 342 as well. The partial products are not as nice here to subtract, but it does work. However, there is another reason to go with the first route — it is a consistent method that works for all products. And with consistent patterns in mathematics, we can create algorithms. In this case, it leads to the use of the distributive property in algebra: $(a − b) × (a − b) = (a × a) − (a × b) − (b × a) + (b × b)$. This algorithm works with variables, variables and numbers and just with numbers. In this case it looks like the following: $18 × 19 = (20 − 2) × (20 − 1) = (20 × 20) − (20 × 1) − (20 × 2) + 2 = 400 − 20 − 40 + 2 = 342$.

The power of using the compensation strategy for multiplying numbers along with the area model and careful notation is threefold. First, we are using important properties, the additive identity and additive inverse properties, when we compensate. Second, we are building flexible thinking with numbers. And third, we are introducing important algebraic thinking along with reasons for why a method works. It can be very helpful for students to explore a strategy such as compensating both factors only to then determine it is not a strategy they would use frequently. Thus, they understand how it is possible, but not necessarily efficient, to use a strategy that is a valuable learning goal beyond just fluency development.

32 × 24

For 32 times 24, the benchmark strategy seems to be easiest. In this case, we can either use the ratio table to keep track of the partial products or we can list them.

I can decompose the 24 into 10, 10, and 4. I know 32 times 10 is 320. Then, I can use the doubles strategy to get 640. Then, I multiply 32 times 4, for which I use a doubles strategy: 32 times 2 is 64 and 64 times 2 is 128. Then I compose 640 and 128 to get 728. So, 32 × 24 is 728.

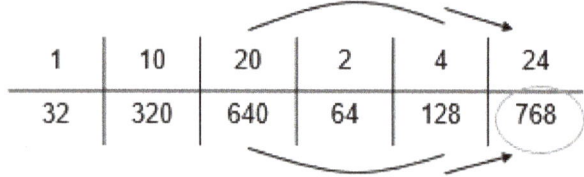

32 × 10 = 320

32 × 20 = 640

32 × 4 = 128

32 × 24 = 728

29 × 42

For 29 times 42 we can compensate 29 to 30 and decompose 42 into a 40 and 2.

If I compensate, I will first decompose 42 into a 40 and 2 and then multiply 30 times 40 to get 1200 and 30 times 2 to get 60. I then know that 30 times 42 is 1260. I compensated 1 too many 42s, so I need to subtract 42 from 1260. I take 40 away to get 1220 and then 2 more to get 1218. That means 29 times 42 is 1218.

30 × 40 = 1200

30 × 2 = 60

30 × 42 = 1260

− (1 × 42) = −42

1260 − 40 = 1220 and 1220 − 2 = 1218

29 × 40 = 1218

Or I could 42 thirty times and then compensate. I know ten 42s are 420 and then I multiply 400 by 3 to get 1200 and 20 times 3 to get 60. These compose to 1260. Then I need to compensate by subtracting the extra 42, to get 1218. So, 29 times 42 is 1218.

1	10	30	29
42	420	1260	1218

If I use the benchmark strategy, I first decompose the 42 to a 40 and a 2 and then use a doubling strategy to determine 29 times 40. 10 times 29 is 290. 20 times 29 is 580 and 40 times 29 is 1160. Then 29 times 2 is 58. I then decompose the 58 to 40 wand 18 to add them to 1160, which is 1200 and 1218. So, 29 times 42 is 1218.

29 × 10 = 290

29 × 20 = 580   (I compensated 300 + 300 = 600 and 600 − 10 − 10 = 580.)

29 × 40 = 1160   (I decomposed 580 into 500 and 80. Then 500 + 500 = 1000 and
                     80 + 80 = 160. 1000 + 160 = 1160)

29 × 2 = 58

1160 + 58 = 1218   (1160 + 40 = 1200 and 1200 + 18 = 1218)

29 × 42 = 1218

Both strategies appeared to be about the same number of steps and mental energy. But the second compensation strategy seemed to be the most efficient.

47 × 58

These numbers are larger, and the ones are close to the next benchmark number of ten. I think compensating both might be the most efficient strategy. 50 times 60 is 3000. Now, I need to subtract three (50 − 47) 60s, which is 180. And I need to subtract two (60 − 58) 50s, which is 100. So, 3000 minus 280 is 2720. Then I need to add the overlap back which is 3 times 2, or 6. 2720 plus 6 is 2726. So, 47 times 58 is 2726.

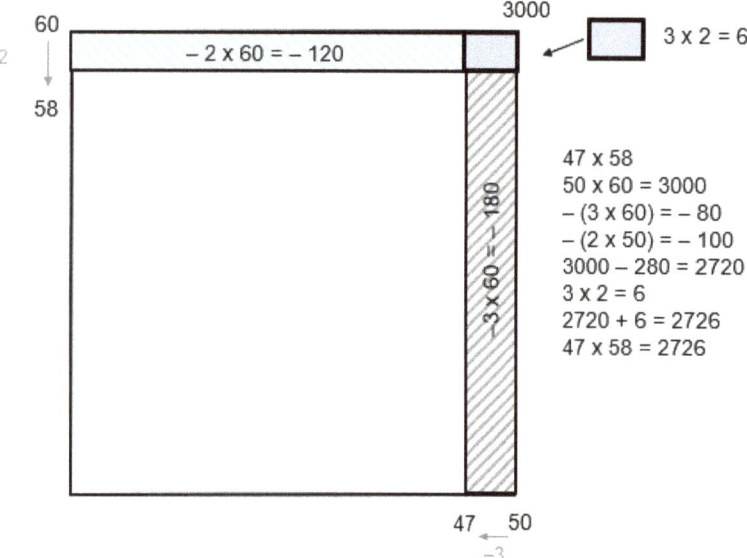

We can also use the benchmark strategy and decompose one of the factors. Let's use the ratio table to keep track of iterating 58 forty-seven times. I double 58 to get 116. Then, I find ten 58s, which is 580. Then, I take half of 10 to get five 58s which is 290. Then I can compose the products of two and five to get 7 and 116 and 290 to get 406. Now that I found seven 58s, I need to find 40. I use a double strategy to get twenty 58s which is 580 plus 580 or 1160. Then I double twenty to get 40 and 1160 plus 1160 to get 2320. Then I compose the forty and seven 58s together, which is 406 plus 2320. This gives me 2726. So, 47 times 58 is 2726.

1	2	10	5	7	20	40	47
58	116	580	290	406	1160	2320	2726

Summary

By practicing using the different strategies with larger numbers, students improve their mastery of multiplication facts, place value, and addition and subtraction facts. Encourage students to use the area model when using compensation, to keep track of the partial products and know what to subtract and add. When using the benchmark strategy to decompose a factor and then compose the parts back together, students will find the ratio table an excellent way of keeping track of all the products.

Fluency and flexibility with multi-digit multiplication using non-integer rational numbers

These strategies can be used to build fluency and flexibility with common fractions and smaller decimal products.

It is much easier to work with whole numbers. One approach that enables this is informally referred to as the "ten rule" whereby you shift the decimal point and work with whole numbers. When we multiply a number by 10, it increases by one place value and the decimal point shifts to the right one place value. Conversely, when we divide by 10 this deceases the place value by one place and the decimal point shifts to the left one place value. This method uses the identify property of multiplying a number by 1, which doesn't change the value of the number. It is a compensation strategy similar to adding 0 as we have used previously.

Let's try an example: 0.3×0.4.

$0.3 \times 10 = 3$	Ten rule
$0.4 \times 10 = 4$	Ten rule
$3 \times 4 = 12$	Multiply the whole numbers
$12 \div 10 = 1.2$	Compensate for using the ten-rule
$1.2 \div 10 = 0.12$	Compensate for using the ten-rule
$0.3 \times 4 = 0.12$	

Mathematically, the process looks like this.

$0.3 \times 0.4 =$

$0.3 \times 1 \times 0.4 \times 1 =$

$0.3 \times 10 \div 10 \times 0.4 \times 10 \div 10 =$

$3 \times 4 \div 10 \div 10 =$

$12 \div 10 \div 10 = 0.12$

$0.3 \times 0.4 = .12$

Another approach is to use the benchmark strategy. If we are multiplying 2.4 times 1.7, we could decompose one of the numbers.

2.4 × 1.7

1	.1	.5	.2	.7	1.7
2.4	.24	1.2	.48	1.68	4.08

2.4 × 1 = 2.4

2.4 × 0.5 = 1.2

2.4 × 0.2 = 0.48

2.4+1.2 + 0.48 = 4.08

2.4 × 1.7 = 4.08

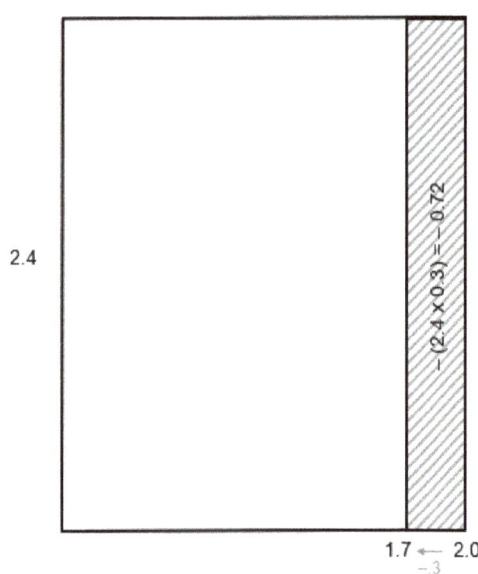

Or we could use a compensation strategy.

2.4 × 1.7

2.4 × 2 = 4.8

− (2.4 × 0.3) = − 0.72

4.8 − 0.72 = 4.08

These strategies can be used for larger decimal numbers as well. But for building fluency and flexibility, it is best to practice with two-digit decimal numbers.

For multiplying fractions, using a decomposing strategy works best. Let's try an example for $2\frac{2}{3} \times 3\frac{3}{4}$.

$2 \times 3 = 6$

$2 \times \frac{3}{4} = \frac{6}{4} = 1\frac{1}{2}$

$\frac{2}{3} \times 3 = \frac{6}{3} = 2$

$\frac{2}{3} \times \frac{3}{4} = \frac{6}{12} = \frac{1}{2}$

$6 + 1\frac{1}{2} + 2 + \frac{1}{2} = 10$

Conclusion

Math fact fluency is essential for learners of all ages. Having the ability to quickly and accurately recall basic math facts and even some common fraction and decimal number operations will allow mental energy to be reserved for more demanding problem-solving tasks. Mathematics is a tremendously complex content area, and fluency skills can unburden students' thought processes and aid their understanding of sophisticated and difficult math topics.

The critical question families and educators should ask is, how should fact fluency be nurtured so that students can be successful? There are several differing views on the matter, and research studies are often too technical to be readily accessible to the majority of teachers and families.

In this book, we have tried to provide clear guidance on the highly effective and balanced methods we have used in Developing Mathematical Thinking Institute partner schools to support thousands of students' fluency. Whether you are a classroom teacher, a member of the school support personnel, or a parent or guardian, it is our sincere hope that you will find implementing our approach to fluency development a promising addition to your work with learners. It is also our desire to make developing math fluency a positive, understandable, and highly connected part of students' math experiences. While this book has primarily discussed fluency, it is helpful to remember that mathematical understanding is not always closely correlated with the speed at which a student works. Ideally, fluency development would be a small part of a more comprehensive and relatable learning experience, in which attaining fluency is viewed as helpful goal but does not represent the entirety of what learners engage in. There is so much to explore and be inspired by in the world of mathematics. The entire staff at DMTI will be deeply grateful if this book helps make fluency one small part of that experience.

www.ingramcontent.com/pod-product-compliance
Lightning Source LLC
Chambersburg PA
CBHW051258110526
44589CB00025B/2867